Method Quarterly

ISSUE 2 | VISIONS

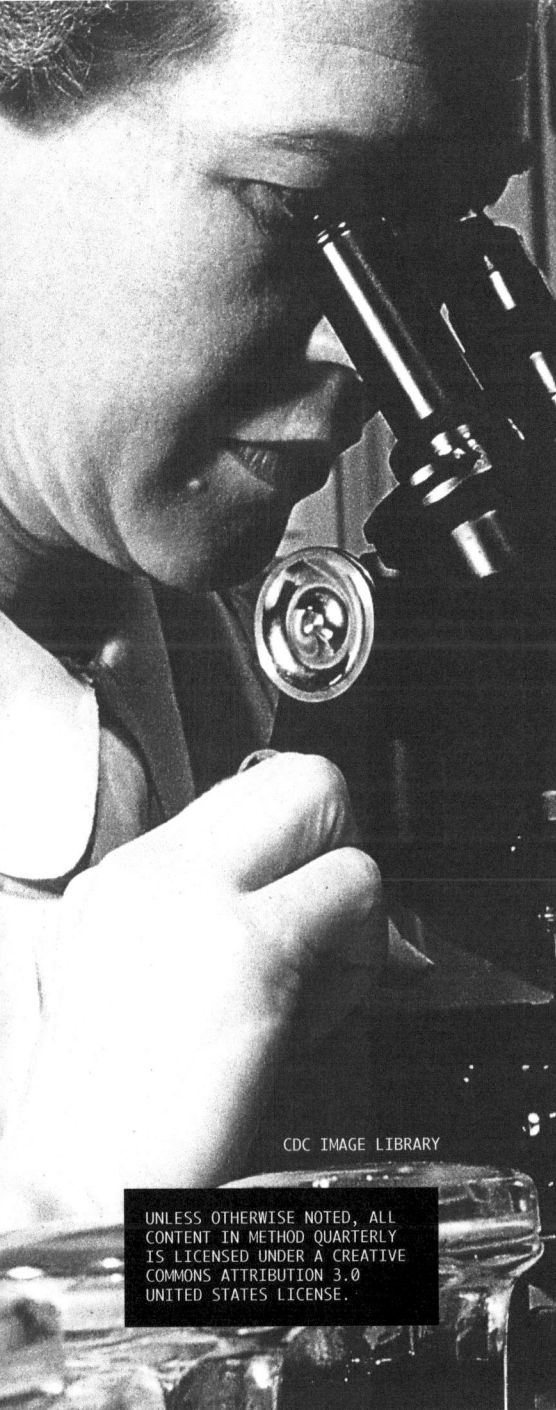

CDC IMAGE LIBRARY

Seeing Science

What does the Higgs boson look like? If you followed any of the news of the boson's announcement in 2013, you probably saw a lot of brightly colored curlicues exploding against a black background. But as Chihwei Yeh notes in this issue of Method Quarterly—on "Visions"—there is actually no boson in these images. The Higgs boson is a perturbation in an energy field, visible to particle physicists only through the statistical ripples it leaves behind.

For the particle physicist, to see is to create. The miles of accelerator tunnels at CERN create subatomic perturbations to make them detectable; CERN's graphic artists then create the images that begin to make these perturbations graspable in the imaginations of people outside of the lab.

Even at a much more familiar scale, scientific vision requires more than just seeing. From ecologists observing natural systems, to astrophysicists scanning the skies, to molecular biologists peering into the eye of a microscope searching for the neon green glint of a fluorescent tagged protein, the work of seeing requires technology, methods, and a vision of what to look for.

Perhaps because of this, scientists are also—often with too little or too much doubt cast—ordained with a sort of visionary foresight. How do scientists consider the future—in their assumptions, forecasts, and hallucinations? Whose visions guide these predictions, and whose don't? And how do they decide what to look for?

In this issue of Method, we look at the ways in which scientists try to observe the world, and how disputes and changes over the ways we choose to see things can in turn alter how we envision the future. We have the astronomers at the Vatican observatory, who—free from the confines of the grant system—are asking some of the farthest reaching scientific questions about what exists in space. We trace the changing tools x-ray crystallographers used to model the twists, curls, and folds of proteins—and how that in turn transformed the way they viewed these microscopic structures. We examine what a new vision for South African science must do to see the human impacts of disease, beyond just its biochemistry. And we ask what the phenomenon of simultaneous discovery says about the myth of the lone scientific visionary. And much more.

The Editors:
■ Azeen Ghorayshi
■ Christina Agapakis

Image: © Xavier Cortada (with the participation of physicist Pete Markowitz), "In search of the Higgs boson: H -> tau tau", 2013.

Chihwei Yeh

Seeing is Believing: Constructing the Higgs Boson

The Higgs boson isn't really a 'thing' in the way that a non-particle physicist might understand the term. How do physicists 'see' it, and how do they negotiate its public image?

In 2013, the discovery of the Higgs boson at CERN (the European Organization for Nuclear Research) caused an international press frenzy. Magazines and newspapers published dozens of stunning images to illustrate the news of the discovery to a general audience. But there was one important problem: none of these images really displayed the Higgs boson.

This came up during the Q&A session following a public lecture given by a particle physicist I'll call Angelica* at the National Museum of Scotland. A non-physicist member of the audience asked, "Where is the Higgs boson in this image?"

To this seemingly simple question, the physicist replied, "There is actually no Higgs boson in it, but only its footprint."

■ ■ ■

The problem boils down to a situation that's pervasive across science communication, which often relies on images to illustrate and represent complex scientific concepts. Images are powerful tools: they can attract people's attention and provide intuitive truth much more easily than words can. The twisted ladder of DNA has become a

---

*because this article is drawn from research under human subjects protection, interview subjects have been given pseudonyms.

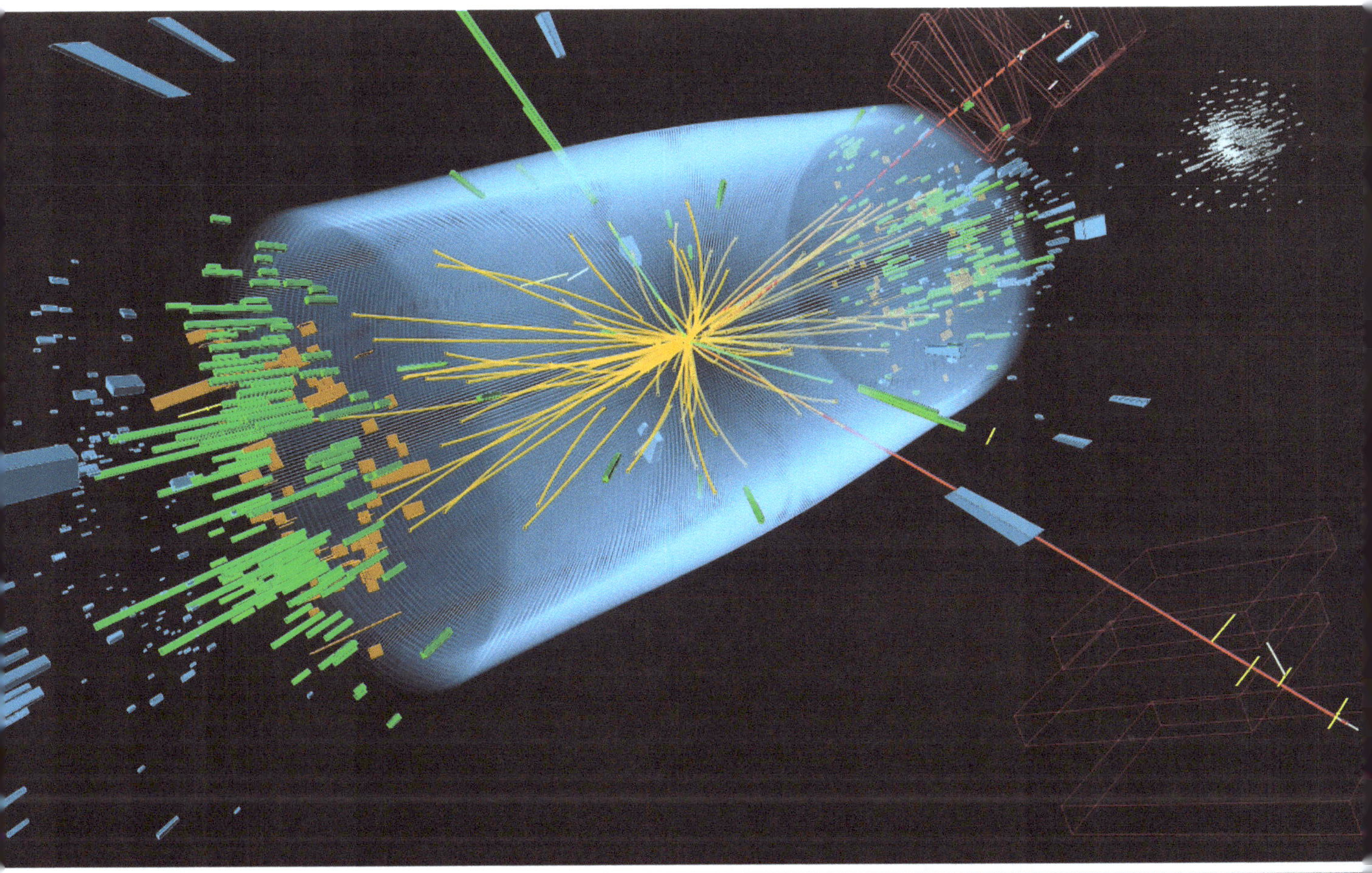

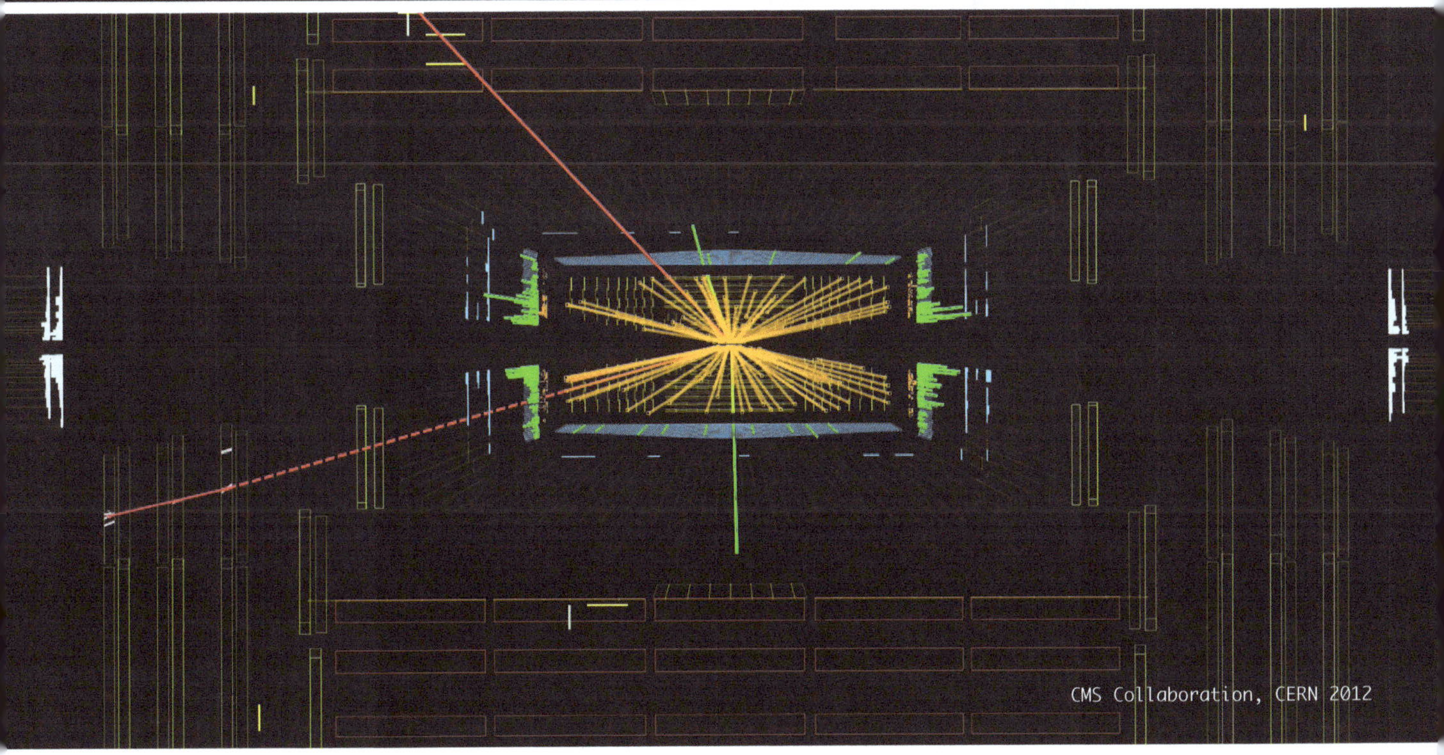

CMS Collaboration, CERN 2012

symbol of the life sciences and life itself. Images of Earth from space, with its recognizable blue oceans, green continents, and swirly white clouds, represent our home planet, and have become powerful symbols for environmental stewardship. But, in seeing these as scientific, we tend to forget that these images are also social products, deliberately manipulated to convey a story or a subjective opinion—and often reflect the vision and culture of the scientific community.

A lot of invisible work goes into turning scientific data into visualizations for public consumption. Most images of a blue marble-like Earth aren't captured by a single camera shot. They're patchworks—mosaics fabricated by NASA from numerous photos of small parts of the Earth. If the Earth is too big for a single photograph, DNA is much too small for photographic treatment. Instead, visual artists and designers create images of the iconic double helix, selling them to multimedia companies as stock photos. The resemblance between an image and reality is actually a product of imagination. Images bound with scientific discourses and activities are not unbiased records of natural phenomena, but are instead artifacts and commodities representing—and often selling—certain ideas about nature.

In the case of the Higgs boson, while the audience expects to 'see' the physical form of the particle, this search is in vain. As the particle physicist points out, it is impossible to actually visualize the subatomic particles. But what do these images mean without an observable Higgs boson? How do they do the work of science communication? Why do science communicators produce such images for the public—and are these images useful inside the scientific community as well?

The Higgs Boson Is an Invisible Field

On the press preview day of the Collider exhibition in the Science Museum London, a journalist asked Peter Higgs about how he himself visualizes the Higgs boson. Higgs responded that he doesn't visualize it at all.

That's because the Higgs boson isn't really a 'thing' in the way that a non-particle physicist might understand the term. Rather, it is a perturbation, a ripple in an energy field. In 1964, Peter Higgs proposed that fundamental particles get their mass by interacting with an ever-present energy field. To prove the existence of the Higgs field, it has to be excited to create detectable ripples. If the Higgs field is an invisible sea, the Higgs boson is a wave on the surface that requires very expensive equipment to be able to 'see'.

It took almost fifty years and many billions of dollars to prove that this field and this mass-giving mechanism exist, because it's extremely difficult to produce these ripples—the Higgs boson—inside the Large Hadron Collider (LHC). And once it appears, it decays immediately, so particle physicists can only record the footprint of the collision. By analyzing the decay of this footprint, the physicists were able to infer that the Higgs field exists.

But the journalist's question to the particle physicist was far from naïve. It demonstrates the challenge of understanding the details of particle physics without knowing the language, assumptions, and methods of the discipline. And, importantly, it also reflects how often the Higgs boson is strategically—and incorrectly—visualized to explain and represent the field's knowledge. This is what science writers Jack Cohen and Ian Stewart have referred to as the 'lie-to-children.' To help make the leap from the LHC to the public, science communicators use sensational images and employ active verbs like 'hunt' and 'look for' to attract the lay audience's attention to this otherwise obscure subject matter. The metaphor of a seeable and chaseable boson is intended as the first rung on the knowledge ladder for more engagement and interaction with the public.

Images for the Public

When I interviewed Angelica she told me that the images she used in her presentation were produced after she and other particle physicists found the evidence to prove the existence of the Higgs boson. The images therefore weren't useful evidence for the particle physicists at all. They were created entirely to serve the purposes of public communication.

When creating these images, numeric information in particular is deliberately eliminated to generate a 'photograph-like' feeling. The CERN scientists used different images in different contexts—referring to the statistical bars and charts to identify the evidence and to communicate the discovery within the scientific community, while showing the graphic images to a lay audience that they imagined as being afraid of numbers and complex physics.

The PhD students in the laboratories are often the artists behind these representations for public consumption; they import the collision data into CERN's self-developed graphics application, and select images that are sleek enough to be displayed. Once these images have been vetted by the the members of CERN, they are uploaded to CERN's document server and are free for use by journalists and other science communicators worldwide.

Despite the fact that these images are post-production artifacts, Angelica explains that these images can work as 'photographs' for a general audience, authentically reflecting what was happening during the experiments. Her idea reflects the widespread belief that photographs are representations of truth.

In her public lectures, along with a basic introduction of particle physics, she uses the photograph-like images as an invitation to the lay audience to virtually witness the eureka moment of the Higgs boson discovery. In *Leviathan and the Air-Pump: Hobbes, Boyle, and the Experimental Life*, Steven Shapin and Simon Schaffer note the power of producing such 'virtual witnessing' technology. It has a literary and social function, and thus can transcend boundaries. It can travel from the experiment to the public sphere to multiply witnesses, beliefs, and—most importantly—trust.

Aesthetics in the Invisible Field

Beyond simple explanation, these photograph-like images provide the flexibility and mobility for further interpretation outside of the particle physics community, for example in enabling artists and designers to add on 'more' layers of aesthetics and imagination to the discovery of the Higgs boson. Particle physicists and museum curators work closely with artists and designers in order to generate an emotional path for the lay public to care about basic scientific research. When working to stimulate

Plot of the transverse mass distributions for events passing the full selection of the H→WW*→lvlv analysis. For the particle physicists at CERN, this is what the Higgs boson look like.

Image: CERN 2013.

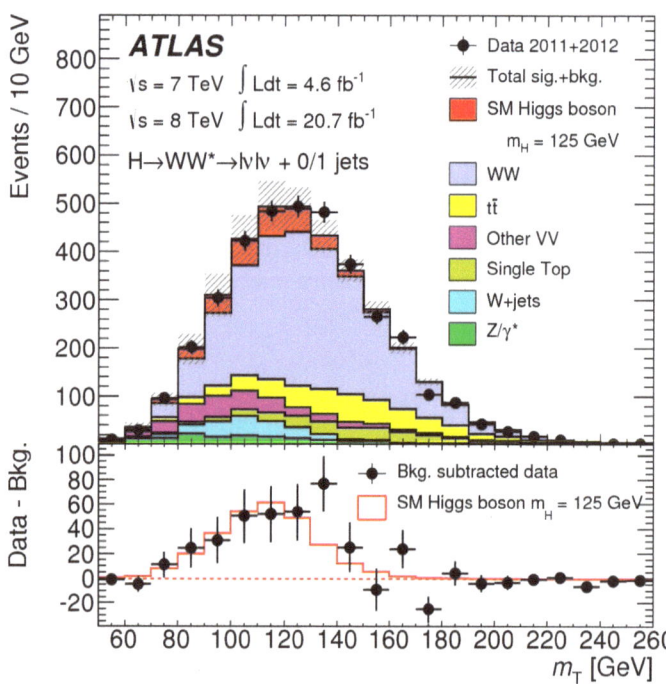

© Xavier Cortada (with the participation of physicist Pete Markowitz), "In search of the Higgs boson", 2013.

public engagement with particle physics, public communication practitioners use aesthetic tools to provide embodied experiences, rather than simply displaying or representing scientific evidence.

In studying the images of the discovery of the Higgs boson used in public communication settings alongside their cross-boundary trajectories, I found that scientific fidelity was not the priority. Instead, both the photograph-like images and the artistic images are stylish representations; they create culture while interpreting the culture of science. Often, they create representations that envision human progress in scientific research, and call for more support of the research.

This creates a paradox: using low-information images to represent scientific authority can often be a strategic choice in science communication. Scientists, artists and public communicators add in colors, shapes, and imagination to make science more compelling. In this way, the Higgs boson—which decays a practically instantaneous 10^{-22} seconds after it's created—has been turned into a space for creative visual representations. Its invisibility allows for the sensational and emotional work that follows. ∎

Chihwei Yeh is a doctoral student in the Science Studies Division at the University of Edinburgh, interested in the social life of particle physics.

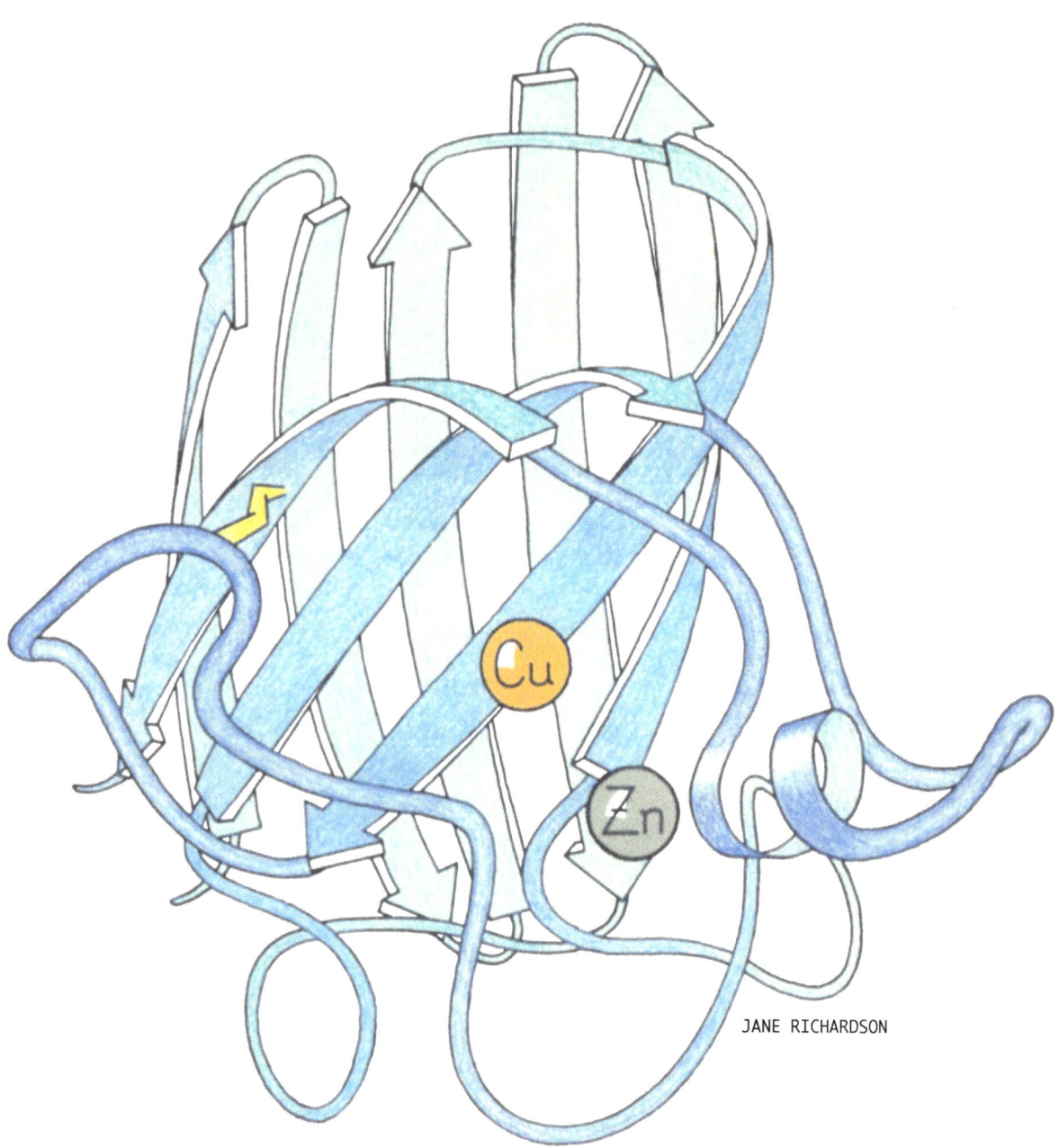

JANE RICHARDSON

Diana Crow

Atom by Atom: Building Protein Models

How we learned to see the folds, twists, and curls of proteins

Before there were computer programs for building 3D molecular models, models of protein structures had to be built by hand. Biochemists in the 1960s and 1970s built three-dimensional representations of their molecules by bending wires into the shapes of their proteins' backbones and adding plastic shapes to represent the atoms. Others used styrofoam sheets and glue.

These models made from wire and plastic inspired the aesthetics of their modern computer graphics counterparts, but were difficult to share with other researchers. Grainy 2D photos, which the journals almost always printed in black-and-white, didn't do the proteins' intricate folds and lengthy amino acid chains justice. And although the journals often printed the entire chemical structure of a newly discovered protein—down to every last hydrogen bond—it was still difficult to discern the similarities and differences between structures.

Jane Richardson of Duke University had read the NIH papers about the structure of immunoglobulin when they first came out in the late 1970s. She and her husband Dave Richardson had spent years poring over x-ray crystallographic images of proteins in order to figure out the proteins' biochemical structure, and the protein they were studying at the time was an antioxidant called superoxide dismutase (2SOD). The Richardsons had noticed from the papers that 2SOD and immunoglobulin shared some common features, but that was typical.

Proteins are chains of smaller chemicals called amino acids. They all have a 'backbone' of interlocking carbon, nitrogen, and oxygen atoms that hold the amino acid chain together, and most proteins have pretty similar ratios of carbon to nitrogen. What sets each protein apart from ananother is the sequential order of amino acids, which controls how the protein folds itself from a one-dimensional chain into a three-dimensional structure. Chemical tests can give you a pretty good idea of how many carbon, nitrogen, and oxygen atoms are in a protein, and genetic tests can tell you the order of the amino acids in the chain, but to figure out the unique folding pattern, you have to obtain an extremely close-up image of the protein. That image can't be a traditional photograph; at the atomic scale of twisting protein chains, the wavelength of light is much too big to actually see anything. To get an image of a protein, you have to shoot it with short-wavelength x-rays and measure the rays that bounce back at you off the protein's atoms.

These images are blurry and difficult to interpret at best. If DNA base pairs are like letters in an alphabet, then proteins are like Shakespeare plays that have been folded into three-dimensional word finds. The motifs that were easy to spot in the genetic data might be folded behind another amino acid group, curled up around something, or interacting with the outer surface of another protein in an unanticipated way. It gets very confusing very quickly.

Structural biology is painstaking, eye-strain-inducing, time-consuming work, and for most of the twentieth century, structural biologists were stuck trying to convey those structures through drawings of chemical bonds, x-ray crystallography data, and black-and-white white photographs of models because those were the types of images that could be printed on journal pages.

So despite reading the immunoglobulin structure papers and noticing a few similarities, the Richardsons didn't fully realize the depth of immunoglobulin's resemblance to 2SOD until they saw the wire model at a molecular biology conference. " I was carrying the [wire model of 2SOD] and a friend who works on immunoglobulin structures was carrying a model of that, and we ran into each other in the doorway," said Jane. "And we just sort of stared at each other's models and said, 'My God! They're the same!'"

The Richardsons and their NIH colleague David Davies spent the better part of the session turning and comparing the two models from every possible angle. The topologies of 2SOD and immunoglobulin were dead-ringers for each other.

Nowadays, researchers can perform the physical analysis the Richardsons and Davies did by comparing wire physical models via computer programs. Not only did these programs not exist when the Richardsons first started in protein biology, but researchers hadn't even settled on an effective method for illustrating which parts of a protein were in front of other parts in the 2D images.

One method the Richardsons used was to draw a series of two-dimensional cross-sections of

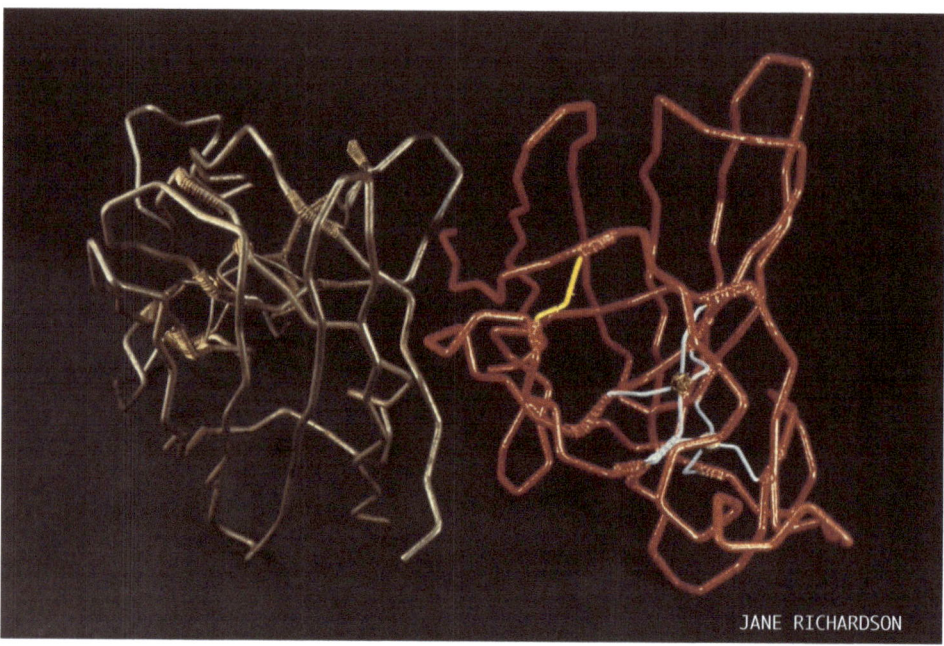

JANE RICHARDSON

the hypothesized protein structure, similar to the way that blueprints include a different 2D floor plan for each section of the building. But those diagrams failed to capture the fact that atoms within proteins aren't organized into floor-like layers; the protein's backbone is twisted into helices or ruffles and ridges, winding its way in and out of each two-dimensional slice.

■ ■ ■

When Jane and Dave Richardson met as undergrads in the early 1960s, the very first protein structures were just debuting. Dave, who was a chemist by training, was immediately fascinated by these elaborate little molecules living inside cells. Jane, who had double majored in philosophy and astronomy, took a bit longer to warm up to them. Many of the chemists at MIT, where Dave began his graduate studies, felt the same. Studying proteins was too time-consuming and highly unlikely to yield useful results.

"My fellow students were saying, 'This is silly; We're never going to get anywhere'—well actually, the professors told me that. I was still excited about [protein structure]," said Dave.

Meanwhile, Jane was working on her master's in philosophy at Harvard but finding

that she missed working on scientific projects. As a teen, she had garnered national attention as a high school astronomer when she placed third in the prestigious Westinghouse Science Talent Search. As an undergrad, she continued studying math and physics before eventually deciding to major in philosophy. But when she moved to Harvard to complete her master's, something didn't quite click. "They didn't focus on the things I was interested in," Jane said. "So I spent a lot of time in the botany department learning about plants."

After completing her botany-infused philosophy master's, Jane ended up getting a job as a technician in the lab where Dave was doing research in as a grad student. The two began working together on the structure of a protein called Staphylococcal nuclease.

"We started out as amateurs. Most of the early structures were being done by the big labs in England and some of the big labs in the US. And we weren't connected with any of those people," said Jane. "It was kind of fun. We'd end up reverse engineering the methodology and visiting the library twice a year to read up on papers. This was such a longshot for the chemist we were working for that he mostly left us alone, which was also nice."

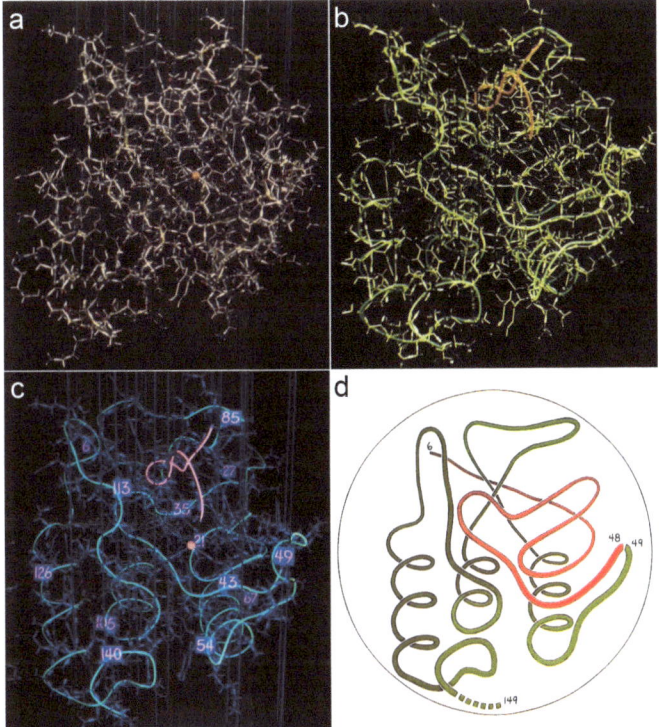

How UV Paint Led to Ribbon Diagams:

Figure A shows a brass model of Staph nuclease under ordinary room light. Figure B combines room light and UV light, highlighting the backbone's place in the overall structure. Figure C is a photo of the model under UV light only. Figure D is one of Jane Richardson's sketches of Staph nuclease, based on observing the model under UV light.

Image: Jane Richardson

The two spent a lot of nights trying to draw proteins on the blackboard and poring over X-ray crystallography data. After the better part of a decade, in 1971, they finally published their solution to the structure of staph nuclease. It was only the 12th protein structure that had ever been solved.

After Dave wrapped up his PhD in 1970, the Richardsons moved to Duke and started their own lab. The lab was technically headed by Dave, but Jane continued working with x-ray crystallography data. In many ways, not being a tenured professor gave her a lot more freedom to work on science without interference.

"I think my philosophic training is very useful. The basis of that is to question everything. Also not having been trained in the same sets of preconceptions as everybody else, that tends to be easier to do," said Jane. "I often had the advantage of being an outsider so I didn't mind as much as most people do sounding dumb. But it gets hard after this many years to keep up that sense that I really am an outsider."

■ ■ ■

Collecting the x-ray crystallography data that allows structural biochemists to calculate where atoms are in relation to each other is a long, laborious process that can take four to five years. Proteins are slippery by nature. It's very hard to get them to line up in a stable crystal. "Growing protein crystals was then, and still is to some extent, a black art," the Richardsons wrote in an email. It could take years to grow a protein crystal large enough to yield reliable data, and processing the diffraction data on 1970s computers was also incredibly time-consuming. But without the high res images that come out of crystallography labs, the biochemists can't do their part of the study; solving protein structures can therefore mean doing a lot of waiting while the crystallography happens, and Jane spent much of the early 70s—while the Richardsons were waiting for the crystallography data about 2SOD—trying to come up with a foolproof way to draw the structures.

"I was trying to find ways to compare other people's structures," Jane said. "The entire first year or two [at Duke] was just learning how to do those drawings. So it was very gradual. I'm not an artist. I can't draw other things all that well."

Since proteins are too small to be photographed directly, representations of them are often based off of earlier representations. One strategy their lab used was to paint the backbone of the wire model with a special paint that glowed under UV light. With the lights switched off, they took photos, and those photos ended up being a key inspiration for Jane's later ribbon drawings of protein backbones. Focusing just on the backbone, it became a lot easier to show the folds and curls of the protein. But it was still not easy.

"A lot of good illustration is, as you probably are aware, taking things away so that you're left with the essential picture," said Dave. "The issue, of course, is that you might cut off something vital and that it might never be recovered."

"The big deal is to really look at the drawing critically and ask whether it really shows what you mean it to show," Jane said. "You have to try to forget all of what you know and try pretend that it's new."

This was one area where Jane's ability to focus her attention on one subject and let the rest of the world fall away really paid dividends. After spending years meticulously diagramming existing protein structures, Jane began to realize that comparing the ways proteins folded could allow researchers to infer evolutionary and functional relationships. In the late 1970s, she began identifying recurring structural motifs in proteins, which culminated with the 1981 publication of "The Anatomy and Taxonomy of Proteins," a mega-review that combined analyses of existing structures with some of the Richardsons' newest research. Although her diagrams had been becoming more and more

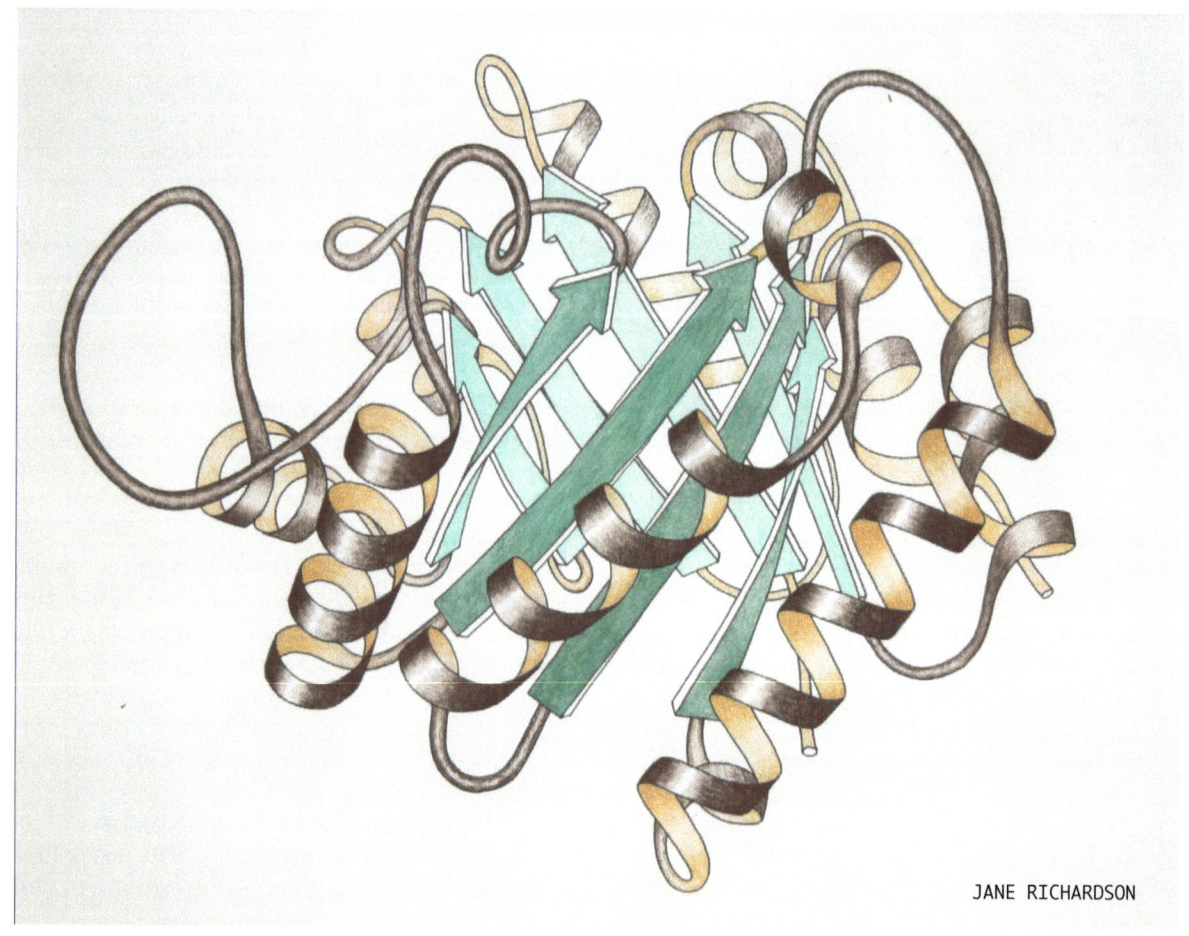

JANE RICHARDSON

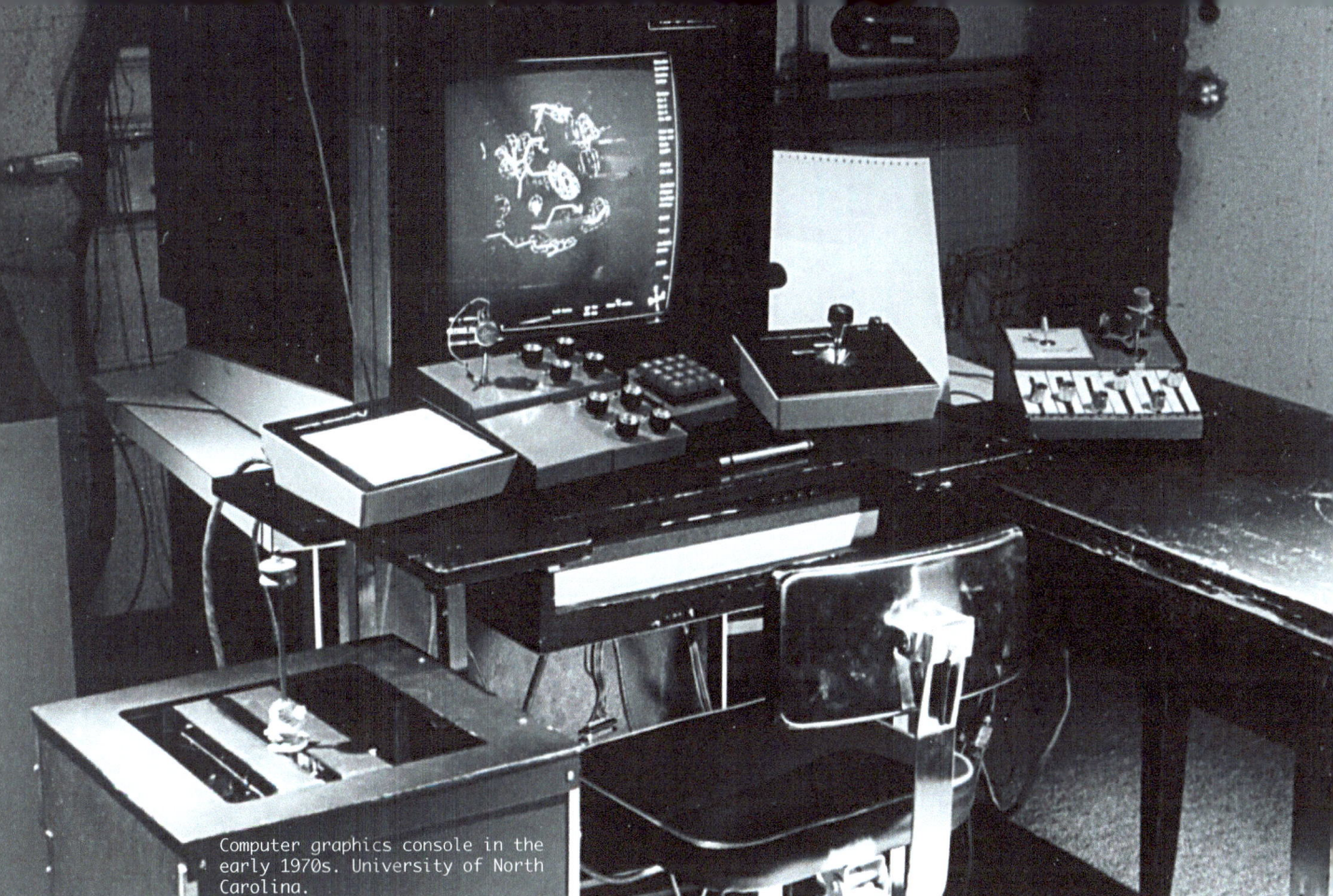

Computer graphics console in the early 1970s. University of North Carolina.

like the now-familiar ribbon drawings, most people in the field point to "The Anatomy and Taxonomy of Proteins" as the ribbon diagrams' official debut. "It was almost one of those things where people said, 'Oh! Why weren't we doing that before?' And it got adopted almost immediately," Jane said.

But the Richardsons also made a concerted effort to help the ribbon diagrams catch on by drawing images of the various recurring structural motifs that Jane had noticed and distributing those images to other researchers for use in slides. By 1985, the ribbon drawings were so widely recognized that the MacArthur Foundation gave Jane a MacArthur Genius Fellowship in order to continue her work. But Duke still hadn't made her a tenured professor.

■ ■ ■

In the early 1980s, the Richardson lab shifted gears slightly and began working on computer-based ways to diagram proteins.

In 1992, they teamed up with the newly launched Protein Science journal to launch Kinemage graphics via a computer program called Mage, which produced rotating 3D images of a protein that could be displayed on Apple desktop computers. Protein Science distributed the first kinemages as article supplements contained on a floppy disk that arrived with each issue.

"Sometimes the authors would make their supplements," said Jane. "But for each issue, I would go through and pick at least a few things that I thought could be illustrated really well, make a kinemage, and then go back and forth with the author."

At first, the kinemages were the scientific equivalent of doorbusters: "When we went to the meeting—it would have been in 1991 probably—we never got to any of the sessions that year, because there was a crowd out where we were showing this, and it was usually 10 people deep," Jane said. But after a while, people got used to them. Other competing programs emerged. Plus

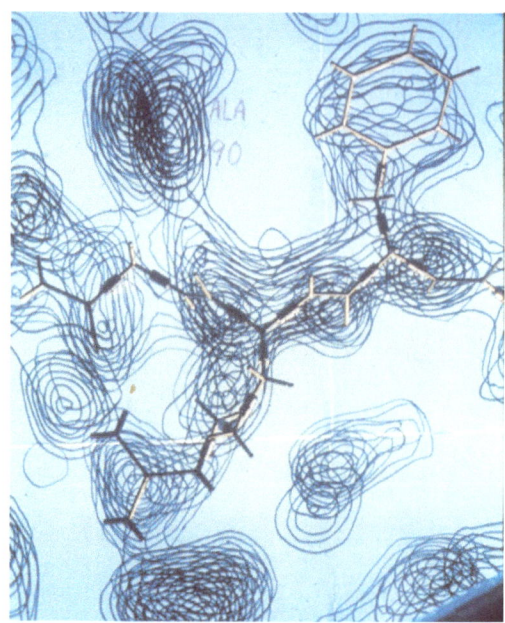

A contour drawing showing the shape of Staph nuclease. Drawing contours onto a clear piece of plastic or glass was one technique the Richardsons used before developing the ribbon diagrams. Image: Jane Richardson.

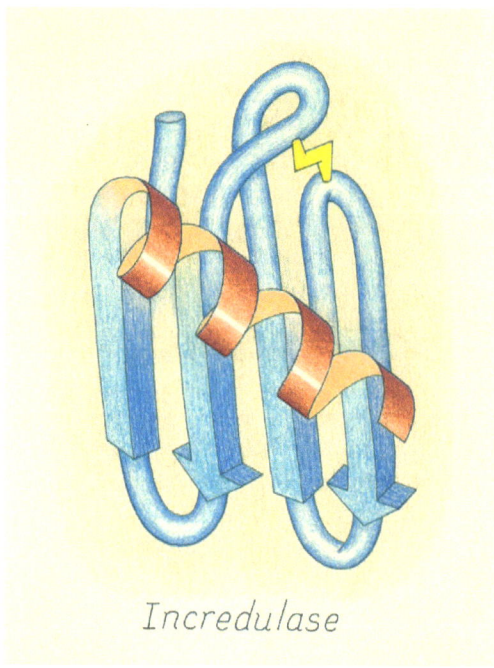

Incredulase: The Model is Not the Molecule. Escher-style ribbon drawing by Jane Richardson.

the demand for 3D graphics got so high that the Richardson couldn't keep up with it. Journal editors began hiring people specifically to work with scientists on graphics and the Richardsons went back to developing new computer programs for studying structure.

Since the early 1990s, the computerized protein visualization field has gotten bigger and more diversified, but the Richardsons have remained key players. In addition to releasing revamped versions of Kinemage, they also developed a program that allows biochemists to identify and diagram the areas of proteins that bind to other proteins or biomolecules (aka the "contact surfaces") and a program called MolProbity that allows researchers to double check their proposed structures for errors and/or unlikely formations. They've also remained active contributors to the Protein Data Bank and are now working on software that predicts the structures of folded RNA fragments.

"We tend to go off on tangents that no one else is doing," said Jane. "Also it tends to be more productive because we've never really liked going to the hottest most competitive thing where you know somebody's going to come up with the answer—if it's possible—even if you don't. It's easier to go off in a direction where maybe nobody else would do it, at least any time soon."

In 1991, Duke finally granted Jane tenure. (That same year, she was elected to the National Academy of Sciences and to the American Academy of Arts and Sciences.) Ever since, the pair has been co-heading the lab, teaching students, and advocating for an approach to visualization that favors interactivity and comprehensibility over aesthetic complexity.

"When we first developed Dave's computer graphics tools, we were doing it for presentation. So we were thinking about it as illustrating journal articles and for teaching," said Jane. "But we discovered that it was what we needed in order to describe the structures to ourselves." ■

Diana Crow is a freelance journalist based out of Boston, MA. She holds a B.A. in biology and spends a lot of time crashing biophysics lectures.

Grayson Clary

The Pope's Planetologists

Doing science at the Vatican Observatory

Brother Guy Consolmagno is very Catholic. His faith is catholic too, insofar as the word literally means universal. And it would be just fine, the Vatican astronomer says clear and often, for alien life to share in it. As he told *The Guardian* in 2010, "Any entity—no matter how many tentacles it has—has a soul." Asked routinely whether he would baptize an extraterrestrial, Consolmagno's habit is to demur: "Only if she asks."

A polemical cottage industry runs on religion, naturalism, and the war between their houses. All the same, many men and women worship on weekends and return, come Monday, to telescopes, spectrometers, and flow cytometers. Often this requires some clever compartmentalization: imagine particles caught in crisscrossed electromagnetic fields of the kind used to isolate antimatter, as if allowing science to touch faith would produce an especially energetic *kaboom*. In that spirit, William Saletan pointed out in *Slate* last year that while Young Earth Creationism can't pass scientific muster, some Young Earth Creationists do leave belief at the laboratory door. But for other researchers, faith is never especially far from the front burner. The most striking are probably those who practice—for a living—both Catholicism and astronomy: the stargazers of the Holy See.

Demographically, astrophysics isn't a notably intuitive choice of field for contemporary worshippers. Physicists and astronomers are, according to a 2009 poll by the Pew Research Center, even less likely to express belief in God than the scientific community at large, edging out geoscientists by a percentage point of incredulity. Still, the Vatican maintains an arm dedicated to astronomical research. Though founded in something of a defensive crouch, designed to illustrate Catholicism's comfort with the sciences, the Vatican Observatory left apologetics behind a long time ago. While much of the public still imagines a Church allergic to physics, its astronomers see the dynamic running in the opposite direction. "If we have a mission at the Observatory," Consolmagno told me, "it's to remind our fellow astronomers that this stuff is supposed to be fun."

■■■

Atop Arizona's Mount Graham, the Large Binocular Telescope and the Heinrich Hertz Submillimeter Telescope share a peak with the Vatican Advanced Technology Telescope (VATT), operated by the Jesuit astronomers of *la Specola Vaticana*. The instruments and their respective operators rub shoulders just fine by all accounts, though in an unfortunate coincidence that's fueled more than a little conspiracy theorizing, one of the Large Binocular Telescope's instruments is named Lucifer. The acronym is awkwardly adapted from "**L**arge Binocular Telescope Near-infrared **U**tility with **C**amera and **I**ntegral **F**ield Unit for **E**xtragalactic **R**esearch." A University of Arizona spokesman told *Popular Science* in 2010 that no offense was meant by the name: "Lucifer just sounds cool."

For what it's worth, Lucifer—literally "the morning star"—has always been an astronomically charged figure in the Christian imagination. Milton gave a famously telescopic description of the devil's armor in *Paradise Lost*:

> *his ponderous shield*
> *Ethereal temper, massy, large, and round,*
> *Behind him cast; the broad circumference*
> *Hung on his shoulders like the Moon, whose Orb*
> *Through Optic Glass the Tuscan Artist views…*

This Tuscan, of course, is the astronomer whose 17th-century condemnation by the Catholic Church

Vatican Advanced Technology Telescope

permanently scarred the faith's scientific reputation – Galileo Galilei.

Apocryphally at least, Galileo responded to the Church's geocentric judgment with one of history's more beautiful statements of scientific fact: *eppur si muove*, "and yet it moves," a jab meant in reference to the sun-orbiting Earth. The incident is still something of a sensitive topic for the Vatican Observatory, whose website has a page on the affair responding to such questions as whether Galileo was an atheist (no), whether he was trained by Jesuits (possibly), whether the Church was "anti-science" then (no), and whether the Church stands by its judgment of Galileo now (no). For an institution two millennia in age, the past is never dead, and hardly past. Not until 1992 did the Vatican make formal amends with that particular deceased Italian.

The Observatory—or at least its modern incarnation—was founded by Pope Leo XIII in 1891 to exorcise Galileo's ghost, whose shadow long obscured Catholic contributions to astronomical research. The idea was, Leo wrote in his founding letter *Ut Mysticam*, to show that "the Church and her pastors are not opposed to true and solid science, whether human or divine, but that they embrace it, encourage it, and promote it with the fullest possible dedication." The Observatory would both highlight and further the Church's scientific achievements. These are easy enough to rattle off, though the Observatory's former director Father George Coyne told *The Baltimore Sun* in 2007 that he preferred not to: "Today we don't preach that. We just do our work."

Coyne, whose studies focused on cataclysmic variable stars, emphasized in a phone conversation that the Observatory's workflow doesn't differ much from that of other research institutions (there is, it should go without saying, no papal or biblical veto on research directions). But the backing of Holy Rome does offer certain advantages. For one, the Observatory's staff of 16 Jesuits is largely exempt from what Coyne called "the sociology of doing science," such as the need to wrangle for tenure or compete for promotions, or the pressures of publish-or-perish. As a result, the Observatory enjoys one of the most essential and basic scientific freedoms: the right to do unglamorous research. As Coyne told me, "We don't have to do—what will I say—the 'gee whiz' projects."

Thanks to the financial support of the Vatican, the Observatory also stands unusually independent from ordinary grant cycles and funding streams. "We wind up doing orphaned science," Consolmagno said, "the kind of science everybody knows somebody should be taking care of but that nobody's willing to sponsor." In the case of its meteorite collection—Consolmagno specializes in meteoritics—the Observatory started collecting data just in case someone might eventually want it. Its researchers can also afford to do science slowly. Very slowly, in some cases: as Consolmagno put it, "We've got a guy who's been doing nothing but measurements of peculiar stars for twenty years."

But control of the 1.8-meter Vatican Advanced Technology Telescope, built in 1993, might be the biggest perk. "A telescope of that size and that quality is very much in demand," Coyne said, and the Vatican's astronomers have it almost entirely to themselves. While they parcel out some time out for collaborations with other institutions—the University of Arizona has a 25% share of the VATT's observing time, while the University of Notre Dame has a contract for 20 nights per year—the rest belongs to the Observatory. In astronomy, it's a luxury to point a telescope not knowing what you may find. With their own instrument, on their own time, Consolmagno said, "It's easier to have good luck."

The VATT was a strategic acquisition that came out of the Observatory's longstanding relationship with the University of Arizona, which in the 1980s was developing a technique for mirror construction involving the rotation of liquid glass in a furnace. VATT's mirror was the first built using the approach, called spin-casting, but the university had no particular use for an instrument of that size. They offered it to the Vatican instead, which promised to raise funds for the device that would house it. "The best way to test it," said Coyne, "was to put it into the telescope." And he was more than happy with the mirror's performance: "It's one of the best imaging telescopes, still, in the world."

■■■

Vatican Advanced Technology Telescope

For all its astronomers' enthusiasm, the Vatican Observatory's scientific work tends to fly under the public radar, as does Catholicism's comfort with modern science more broadly. Much of the news media was shocked last October to hear Pope Francis endorse evolution and the Big Bang, yet neither has been especially controversial in the Church for the last half-century, and the latter was theorized in its earliest form by an ordained Catholic priest. Instead, the Observatory catches attention for its more sensational musings about astrophysics—and no subject is more attractive, or more speculative, than astrobiology. When the Vatican starts talking about ET, a tidal wave of headlines follows.

In 2008, it was Observatory Director José Funes telling *L'Osservatore Romano*, "Just as there is a multiplicity of creatures on earth, there can be other beings, even intelligent, created by God. This is not in contrast with our faith because we can't put limits on God's creative freedom." In May of 2014, it was Pope Francis giving a homily in which he mused, "If—for example—tomorrow an expedition of Martians came, and some of them came to us, here… Martians, right? Green, with that long nose and big ears, just like children paint them… And one says, 'But I want to be baptized!' What would happen?"

Considering that the first extra-solar planet was discovered in 1992, the question is surprisingly old. Neil deGrasse Tyson's *Cosmos: A Spacetime Odyssey* opened on the case of Giordano Bruno, a 16th-century friar who dreamed of extraterrestrial messiahs redeeming alien worlds. He was executed, probably less for his cosmic pluralism than because he rejected—among other doctrines—the divinity of Christ. Religious imagination of other worlds is even older still, with more than a few spiritual traditions puzzling over what, precisely, it would mean to encounter life in the great Out There.

C.S. Lewis laid out one of the best-known Christian treatments of the subject in a 1958 essay called "Religion and Rocketry." There were, he pointed out then, at least five questions that needed answering before the impact of extraterrestrial life on Christian faith could be worked through. He asked whether alien life exists, whether it is ensouled, whether it is "fallen" in the way that the doctrine of original sin holds humanity is, whether it was redeemed through Jesus Christ, and whether other methods of redemption are possible. The sticking point, of course, is that intelligent life is still a sample size of one. 57 years later, these questions remain, to say the least, speculative.

When we spoke, Coyne declined to guess at answers. "We're in the early stages of this whole investigation," he said. "Where's the evidence?" Lewis had held that position too, writing, "If I remember rightly, St. Augustine raised a question about the theological position of satyrs, monopods, and other semi-human creatures. He decided it could wait till we knew if there were any. So can this." But Consolmagno, while conceding that he had no more data than anyone else, expressed confidence (and excitement) about the prospect of life elsewhere in the cosmos. Why? "In science," he said, "so often the better story turns out to be the better science." Theorists, religious or not, do typically prefer their theories elegant. And in the case of astrobiology, Consolmagno, said, "A universe where we are utterly alone, a highly unlikely

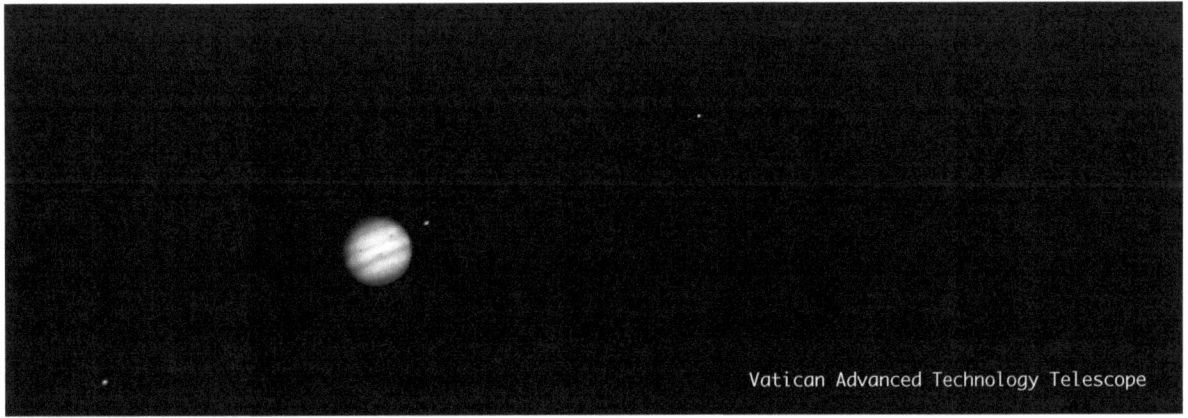

Vatican Advanced Technology Telescope

accident of chemistry, is not as good a story, is not as elegant a universe, is not as beautiful a way that things could be. I'm someone who will always go for more elegance, the better beauty – the better story."

■ ■ ■

Whenever humanity dreams beyond Earth, faith follows. Buzz Aldrin, a Presbyterian, held a small communion in the Apollo Lunar Module. "In the one-sixth gravity of the moon," he later recalled, "the wine curled slowly and gracefully up the side of the cup." At the time, he read John 15:5 – "I am the vine, you are the branches. Whoever remains in me, and I in him, will bear much fruit; for you can do nothing without me."

It's easy to sound unhinged talking about either faith or extraterrestrials. Both habits come, in the public imagination, with special hats. But the Vatican Observatory isn't very concerned on either front. It does its work; sometimes, it wonders evocatively out loud. I heard echoes of *The Imitation Game*—"Sometimes it's the people whom no one imagines anything of, who do the things that no one can imagine"—when Consolmagno explained how his faith gives him the confidence to imagine and speculate. "You're more willing to ask the crazy questions," he said, "if you're not afraid of being thought crazy."

After all, the Church claims a catholic writ; it can hardly be expected to ignore the stars. ■

Grayson Clary lives and works in Washington, DC.

NASA

Cross-section of a douglas fir.
Image: © Henri D. Grissino-
Mayer, The University of
Tennessee, Knoxville.

Meritxell Ramírez-i-Ollé

The Social Life of Climate Science

Scientists depend on relations of trust to produce knowledge about climate change from tree rings

I did not know that scientists could use tree rings as a window into the history of the Earth's climate until the so-called 'Climategate' story broke in late November 2009. In the now familiar tale, mainstream media in the UK and the US reported that emails and other documents from one of the world's leading research centers on climate science—the Climatic Research Unit (CRU)—had been stolen and published online. This private correspondence included conversations among climate scientists working on creating a record of historical temperatures using tree-ring data from forests in the Yamal Peninsula in Siberia and across the Northern Hemisphere. A few bloggers and columnists interpreted the CRU emails as evidence of scientific fraud, and the scandal exploded from there. For months after the leak, the CRU climate scientists were investigated by six university and parliamentary inquiries, which eventually exonerated scientists from allegations of malpractice.

As the Climategate scandal played out in the news, I was a graduate student at the University of Edinburgh, studying how societies both influence and are influenced by science. Steeped as I was in the social realities of science—the politics, the professional commitments, and the relationships that drive laboratory practice—I was initially baffled by the furor generated over the stolen CRU emails. Why was the private correspondence of a few scientists of interest to anyone? What do these emails say that is so controversial? I read a few that were available online, but I couldn't understand what was so polemical about them. Other sociologists of science who analyzed the emails concluded that the CRU emails showed 'scientific business as usual.' The philosopher Bruno Latour sarcastically wrote that the Climategate emails showed:

> **CC** *that the scientific facts of the matter had to be constructed, and by whom? by humans! Squabbling humans assembling data, refining instruments to make the climate speak (instruments! can you believe that!), and spotty data sets (data sets! imagine that…), and those scientists had money problems (grants!) and they*

had to massage, write, correct and rewrite humble texts and articles (what? texts to be written? is science really made of texts, how shocking!)…"

While most sociologists of science were saying 'and so what?' about the stolen CRU emails, I wondered what the climate scientists who had been directly or indirectly involved in Climategate were thinking. So as part of my Master's research, I looked at how individual scientists and scientific institutions reacted to the allegations of scientific fraud. I discovered that, quite paradoxically, they interpreted the emails much in the same way as their critics—as embarrassing deviations from the social demands of a consensual, objective, and open climate science.

But it's not just what climate scientists say that's important, it's what they do—in the field, in the lab, and in conversation with other scientists. The allegations of fraud made against climate scientists during Climategate made it clear that members of the public hold widespread assumptions about what is 'normal' scientific practice. But except for climate scientists themselves, very few people actually know about the process of climate science 'in action'—how scientists build climate models, how they interpret data, and how they interact with other relevant groups in society.

As part of my doctorate, I decided to study how scientists produce knowledge about climate change from trees. I found the website of a tree-ring laboratory in Scotland doing research similar to the CRU scientists at the center of the Climategate scandal. When I first reached out to researchers in the lab (so-called 'dendrochronologists'), they suspected I might be an undercover 'climate skeptic,' intent on exposing the next Climategate scandal. It took them a few months to get to know me well and to understand my research interests.

Since then, I've become a 'temporary lab rat' in Dr. Rob Wilson's tree-ring lab at St Andrews in Scotland. I carried out lab work with Rob's PhD student, Miloš Rydval; I have been out with Rob and his team on fieldwork trips as they collected samples from trees in the Scottish Highlands, and I have followed them to the other side of the world as they presented

The author (left, taking notes) in the field with members of the lab.

their tree-ring based reconstruction of the Scottish climate at a conference in Australia. This day-to-day experience in the lab gave me a firsthand understanding of their methods and tools, as well as a glimpse into their challenges, successes, and failures.

One of the most important activities in the tree-ring lab in St Andrews is tree-ring dating, the key method that dendrochronologists use to determine the calendar date of each tree ring. From there, dendrochronologists work to estimate historical climate; the width of tree rings varies with, among other things, temperature and precipitation. The result of tree-ring dating is a chronology that can be extended back in time beyond the age of a tree by overlapping it with other chronologies generated from archaeological beams and submerged wood. The longest tree-ring chronology created anywhere in the world was built by European researchers using Irish oak trees, currently reaching back about 10,000 years. Long tree-ring chronologies are the basis upon which dendrochronologists have been able to date the remains of the ship at the World Trade Center, and to estimate temperature for times before thermometers were in widespread use, like with the famed 'hockey-stick' graph that puts current global warming into historical context.

The creation of tree-ring chronologies requires that dendrochronologists establish the total number of rings on the wood. However, they insist that tree-ring dating is not about ring-counting. They argue that ring-counting leads to inaccurate dating because it does not take into account anomalous rings. They refer to these anomalies as 'false rings' and 'missing rings.' A false ring is a duplicate-like annual ring that is formed because of unfavorable conditions within a growth year, while a missing ring is an absent ring that has not grown during a year due to sudden cold weather or drought.

To get an accurate dating of the tree-rings, it is crucial that dendrochronologists detect false and missing rings and adjust the count accordingly. How do they manage to see rings that do not exist or ignore the rings that they identify as redundant? Dendrochronologists see the correct tree-rings thanks to the trust

relations they establish with colleagues, which constrain what they perceive on the wood samples. In the lab in St Andrews, these trust relations were constituted in the lab—if not before during fieldwork—where we started preparing the wood. We followed Miloš' instructions so that tree-rings became more visible. We cut, sanded and glued the wood in a way that rings were displayed in line like a bar code with a sequence of wide and narrow rings (image opposite). Once all cores and samples were prepared, we scanned the samples and converted tree-rings into digital images that we could observe on a computer screen.

As the counting begins, the most frequently asked question in the tree-ring lab in St Andrews was, "Is this a ring?" Neophytes like me were often unsure about the exact boundaries of a tree-ring and we sought help from Miloš or Rob while pointing to the problematic ring on a computer screen. Our perception of the wood, and the subsequent classification of a dark band into a 'tree-ring' or 'non-tree-ring', was constituted through, and in turn constituted, relations of trust between new and expert practitioners.

Dendrochronologists make the ultimate decision about whether a tree-ring is a false or missing one by comparing it with rings from nearby trees, the assumption being that these trees would have a similar ring pattern because they had grown in the same environment. Dendrochronologists call this pattern matching 'cross-dating.' In Scotland, we cross-dated rings by measuring their width and using correlation coefficients and a line graph to assess their similarity. In other parts of the world where tree-ring patterns are more easily distinguishable, dendrochronologists cross-date rings by comparing samples side by side. In this way, Miloš often cross-checked his datings—especially of preserved wood—with his doctoral supervisor, Rob. If we found consistent asynchronies between samples, we assumed that they were due to the presence of a false or a missing ring. Generations of dendrochronologists have consistently found that cross-dating works well with a broad range of tree species around the world to the extent that it has become a

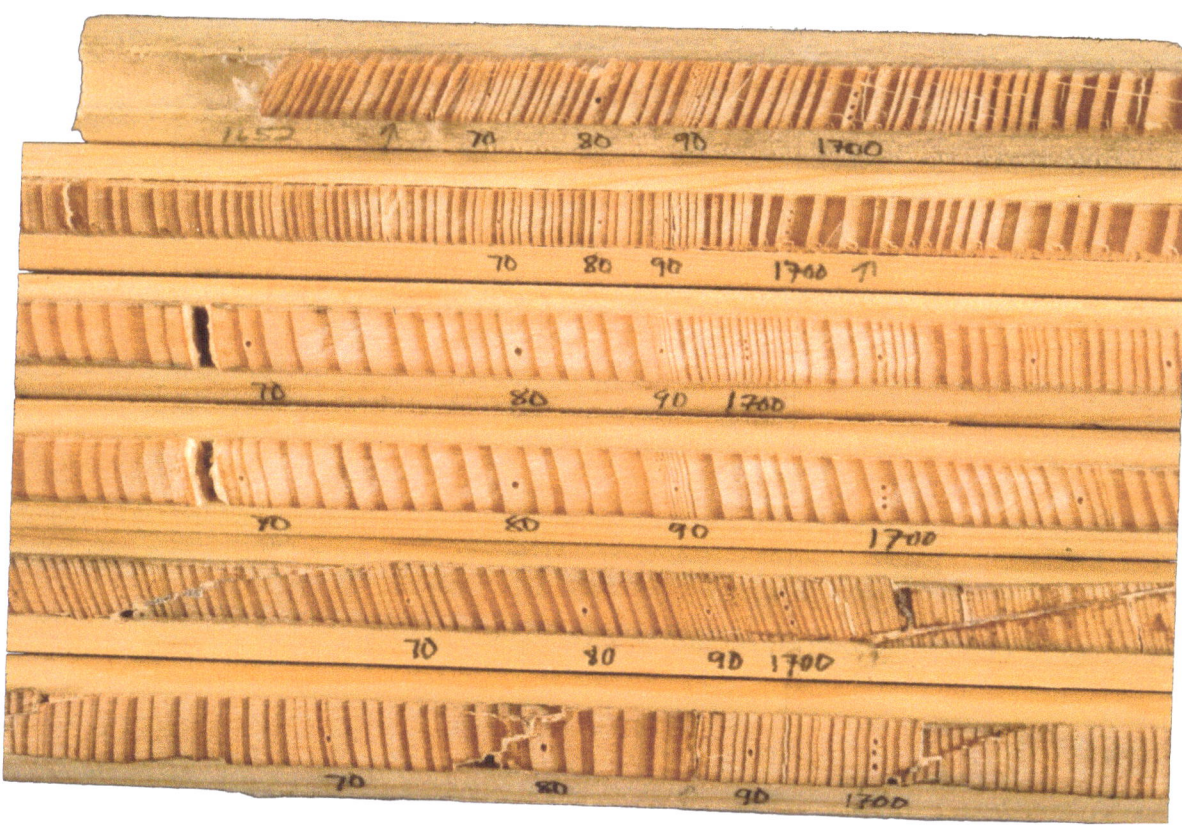

Dendrochronology and tree-ring dating consists in establishing a count of tree-rings that excludes the presence of double and missing rings by pattern-matching. Image: © Henri D. Grissino-Mayer, The University of Tennessee, Knoxville.

'principle' in dendrochronology. In this way, when dendrochronologists work to cross-date rings, they draw on the tradition of knowledge of dendrochronology and, simultaneously, contributes to its existence. A competent dendrochronologist is someone who interacts with members of the community and becomes familiar with the tradition of expectations built up by colleagues so that they see rings in a way that neophytes have not yet learnt to see them.

If you're a scientist, it's probably not surprising to you that learning to see like a scientist is a difficult process that depends on social relationships between students, researchers, and mentors. My hope in doing this research is that recognizing the inextricable—and, I would argue, beneficial—social nature of climate science could help to create a more fruitful conversation between climate scientists and nonscientists. This conversation does not need to exclude mutual criticism, but it has to avoid using social factors as a source of public condemnation. Sociologists of science can't—and perhaps shouldn't—work to prevent future Climategates, but by making the social reality of the lab visible, perhaps we can help rethink the assumptions of science shared by scientists and members of the public alike. Perhaps in the future, as we learn more about how climate scientists learn and do their work, the fact that they negotiate over email how to generate convincing evidence won't be so controversial. ■

Meritxell Ramírez-i-Ollé is a doctoral student in the department of Science, Technology and Innovation Studies at the University of Edinburgh.

Adam Rothstein

The Greatest Thing Since Sliced Bread

Snowclones and the history of the next big thing

RAINER ZENZ

I f one described the computer as one of the most important technologies of the 20th century, it would be tough to disagree. But when we talk about the computer, what exactly are we talking about? Is it the hardware? What about the software, the people who wrote it, or the industry for which software was written? What is different about the history of computers, and the history of electronics more generally?

To Michael Sean Mahoney, a historian of computers, the computer was a model for the difficult practice of history, an example of the problems of defining the scope of research, and the limits of definitions. To talk about the 'inventor' of a technology—or an epochal definition like the 'computer revolution'—is automatically to interact with a preconceived model of history. "'[R]evolution' is an essentially historical concept," Mahoney wrote in *Histories of Computing*. "Even when turning things on their head, one can only define what is new by what is old, and innovation, however imaginative, can only proceed from what exists."

We define the new via the old by reaching across different historical contexts, and finding similarities between different things in different times. Consider comparisons between computers and the Model T during the birth of PCs in the late 1970s and early 1980s. Here's Mahoney again:

❝ *Apple, Atari, and others have boasted of creating the Model T of microcomputers, clearly intending to convey the image of a car in every garage, an automobile that everyone could drive, a machine that reshaped American life. The software engineers who invoke the image of mass production have it inseparably linked in their minds to the automobile and its interchangeable variations on a standard theme."*

Using the Model T to invoke the idea that "this technology will become ubiquitous" is hardly original. Just ten years after the namesake device's invention, Henry Ford himself boasted that the Ford Trimotor aircraft would do for the air what his Model T did for the roads. Of course, the Trimotor failed to have such iconic success, due to obvious differences between the aircraft and automobile markets. Simply to have Ford apply his manufacturing procedures to an aircraft would not be enough to duplicate the success of the automobile, which succeeded as much because of construction materials, financial markets, urbanization, and demographics as it did because of anything attributable to Ford himself. The historical contexts are different. A would not be to B as C was to D, because historical context is rarely transitive.

Today, analogies between technological devices and the Model T are repeated to the point of cliché. With a simple Google search, one can find "the Model T of" aircraft, bicycles,

CHARLES STARRETT

hoverboards, electric cars, driverless cars, payment systems, drones (two different options), satellite launchers, and even the biological mechanics of sharks. The Model T comparison has ceased to be a symbolic definition, and has instead become what's been dubbed a snowclone: the act of comparison itself, repeated until it is noise rather than signal. Bereft of context, the snowclone becomes an idea echo chamber, repeating the sound of the idea until the idea itself is mostly unintelligible.

The snowclone gets its name from a 2004 discussion on the University of Pennsylvania's Language Log, originally utilized to describe the lazy and clichéd journalistic practice of repeating some version of: "If Eskimos have N words for snow, X surely have Y words for Z." Other examples are riffs on the *Alien* film poster tagline "In space, no one can hear you X," or the ubiquitous "X is the new Y." Snowclones are verbal memes, ideas that are formulaic, ultimately repeatable, and meaningful only in the context of that repetition. Even doge, perhaps the *ne plus ultra* of memes, qualifies as a snowclone. "So X, very Y."

There does seem to be something meme-like about the evolution of historical technology analogies into stock phrases. Companies deploy memes for branding purposes all too often these days. There doesn't have to be a carefully considered historical comparison; if you want to sell something to a market segment that has already accepted a particular product, labeling your new product as analogous to that older product is a foot in the door. The familiarity of the meme and of the referenced product makes it repeatable *ad infinitum*, a cookie cutter to shape new products. Today, the iPhone snowclone is almost as ubiquitous as iPhones themselves. "The iPhone of X" works if you want to appeal to iPhone owners, or if you want to get a piece of that ubiquity. There is the iPhone of drones, of vaporizers, of online business platforms, and even of guns.

The iPhone was all but non-existent before 2008, yet six years later is ubiquitous. It has altered our everyday lives in countless ways, but it is impossible to be sure which ways will have a lasting effect. Will the ability to have email updates pushed to our phone still be significant ten years from now? In 50? Will we always claim to have "an app for that"? For how long? When we label a technology as the "iPhone of X", we really aren't saying why X will be significant, because we aren't totally sure about the iPhone, either. We know or desire that there will be some longer effect in the future, but we don't know what that may be. We have a sense that there is a something here in our moment in history, our historical context, that is significant, and yet we cannot say for sure what it is. Snowclones dance around historical context, but that doesn't mean that there is nothing to learn from that dance. The historical specifics of any particular technology may be completely glossed over in the rush to paint a certain marketing picture. The material facts of history are not always considered, but there are important stories being told, all the same. In snowclones we see society's technological desire, writ large.

The technological snowclone speaks to our notion of historical epochs, our sense that new technologies mark a historical transition, unique to their times in a superlative way. In many cases, it doesn't necessarily matter that the Model T and the PC—or for another example, the Wright Flyer and the Saturn V rocket—are drastically different technologies created in completely different contexts. What matters is that each achieved a certain breakthrough, and thus launched the 'automobile age,' the 'computer revolution,' 'the age of flight,' and the 'space age,' respectively. But these epochs also betray a certain insufficiency, a deeper context missed in the symbol. We are still flying in planes, and aircraft are still evolving, even as we launch satellites into orbit. If these ages can be contemporaneous, what does that say about their relationship to each other? What does the 'age of flight' mean, if today even our simplest aircraft are completely incomparable to the Wrights' cloth and wood construction?

Consider also the 'moonshot,' a phrase used today by Google X, the division of Google that explores inventions that would satisfy some larger human dream—not unlike how the Wrights satisfied the urge to fly, and the Apollo missions were the final push of a chain of inquiry stemming back to the earliest astronomers. The phrase means risk, a distant goal, and uncertain payoff outside of a wider, scientific glory. And yet in Google's context,

NASA

the phrase now describes self-driving cars and broadband-by-balloon. While both are certainly technological feats, it is difficult to say whether they would usher in a new era of human history. What does the 'moonshot' mean to us today, when we are still struggling to get back to the moon? Why aren't Google X's 'moonshots' actual, literal moonshots?

Often, while technological snowclones stay much closer to home, they maintain a distance between our current understanding of technology and any broader context—historical, economic, or political. Take, for example, the Walkman and the iPod. These two personal music players have similarities in the way they affect our ability to listen to personal music in public. These particular models so dominated our understanding of the device, that their brand names are now used generically to refer to any similar technological system. On the other hand, if we were to present these two technologies as analogous in their use by consumers, we would also need to draw their secondary effects on the means of music production and distribution into comparison as well. And there might be certain similarities: the ability for consumers to dub their own tapes threatened music industry profits, as did the mp3. But the iPod, through its relationship to the iTunes store, in some ways *saved* the music industry's line of profit from the threat of the mp3 itself, a totally different technology, which is not analogous to the cassette tape. So while these devices are similar, the iPod's relationship to online mp3 sales is a difference so profound that we have dubbed the latter the 'iPod moment,' now used to analogize what iPods did for music consumption to smart watches, payment systems, newspapers,

ED UTHMAN

e-readers, electric cars, and 'thin clients' (whatever they are). And so with all vocal similarities, silent differences are introduced as well.

Of course, the technological snowclone does not necessarily intend a literal analogy, but can be more about what we consider to be 'the spirit' of invention. Consider this Sprint commercial, in which an entire canon of devices are portrayed in a 'domino effect' of history, cascading from one into the next. There is certainly a technological system extending across human history. New inventions set the stage for the next wave of inventions, in the market, in manufacturing, in sub-components, and in human recognition of what is possible. But this domino effect implies analogical succession, a genealogy of technical inheritors. 'Progress' is a network, not a line; the invention of the camera simply does not have anything directly to do with the invention of the cell phone—or at least, no more that it does with the invention of the e-cigarette, which is not pictured in the Sprint commercial for the obvious reason that it not considered culturally significant, part of the same inherited greatness. This Biblical, canonized chain of begetting leads, unsurprisingly, to Sprint's newest smartphone—the descendant of a techno-historical line of David.

Overall, these analogies tend to portray technological objects as flat, 'great gadgets of history'—attributing their importance to the devices themselves, rather than the historical context that facilitated their creation. Consider the "Homebrew Computer Club," a loose association of PC experimenters famed for gathering in the '70s and '80s, at a time when many of them were pioneering concepts in personal computer design. The snowclones of the Homebrew Computer Club can now be found for DIY hardware, quantified self projects, law, biology, micro satellites, fab labs, drones, and robots. Just as Henry Ford did not invent the car when he built the Model T—he was a product of his context, pioneering a production model able to produce cheap cars fast enough to keep up with demand, while continuing to expand his production operations—the Homebrew Computer Club did not invent personal computers, though many members did launch computer firms. A particular Silicon Valley social

milieu is also only so interesting—it is likely that if Apple Computer hadn't produced the iPod and the iPhone, no one today would invoke the company's founders' hobbyist groups in the late 1970s. Certainly, no one speaks about the founders of Osborne Computer and Morrow Designs in such hallowed tones, even though they both took part in the group as well. These are the sorts of gadgets that society prefers—those with single inventors, burst upon the scene in a singular event, that then take the world by storm.

Across these modes of analogy to many, various "greatest things since sliced bread," we interact with sometimes a little bit of historical context, sometimes a lot, and other times none. But what unites them all is that this intention is, in fact, unspoken. They reference history, and yet they are always partially fantasy. It may be asking too much of popular media to have a depth of historical analysis in its tech writing. And yet the entire record of history is often excised in the simple use of any of these snowclone analogies.

But if we can easily see what is left out, what is actually said? When wading into the convoluted process of deciding what happened in history, what was relevant and what mattered, the cognitive biases implicit in our societal storytelling are revealed. 'Ubiquity,' 'change,' 'similarity,' 'epochs,' 'evolution,' and 'invention,' are all stock tropes by which we explain the complicated context of history when we summarize it. These are not just lazy forms of speech and writing, but a window into our sense of time. These are the dreams of our cultural sub-conscious, and the obstacles any explanation of history must avoid. But at the same time, we are stuck with them, because even though it is easy enough to see the ways in which snowclones fail, they are what we reach for immediately. Like a nervous tick or a habit of speech, our historical concepts are part of us. Understanding our continuing technological history must avoid snowclones, but as long as that history is human, it will never fully escape them. ■

Adam Rothstein writes about politics, media, art, and technology wherever he can get a signal. He is on Twitter @interdome, and his new book Drone is out from Bloomsbury Publishing.

Amisha Gadani*

The Femunculus

Male bodies are seen as the default in biology; we must learn to see the female form

In psychology, the homunculus is a visual representation of the sensory and motor cortices. In one version, the body is stretched and distorted over a cross-section of the brain, body parts mapped onto the areas of the brain that control them. In another, the body is represented with the relative sizes of the parts inflated to proportionally represent their relative neural footprint—with oversized hands, lips, and genitals.

Homunculus means "little man," and historically, the sensory and motor homunculi are represented as male figures. In studies of human biology, male bodies are seen as the default, and as a result, medical devices such as artificial hearts have been built with men's bodies in mind—too big for a significant percentage of women. Early stage clinical trials for drugs have also historically excluded women in order to "simplify" the analysis, which has led to drugs that have unknown and even dangerous effects in women due to important differences in dosing and metabolism. Moves to address this problem at the funding level have occurred only recently.

SENSORY HOMUNCULUS

The consequences of the sensory homunculus excluding women are perhaps less dramatic, but suprisingly it wasn't until 2009 that neuroscientists specifically looked at how women's bodies map onto the sensory homunculus, creating the first scientific *femunculus*.

Amisha Gadani is an artist whose work explores biological diversity and evolution through illustration and kinetic fashion design. In a series of drawings based on results from the belated studies of female bodies, she proposes a visual representation of the femunculus, reminding us that there is more to see beyond the scientific default. ■

*Text by the editors

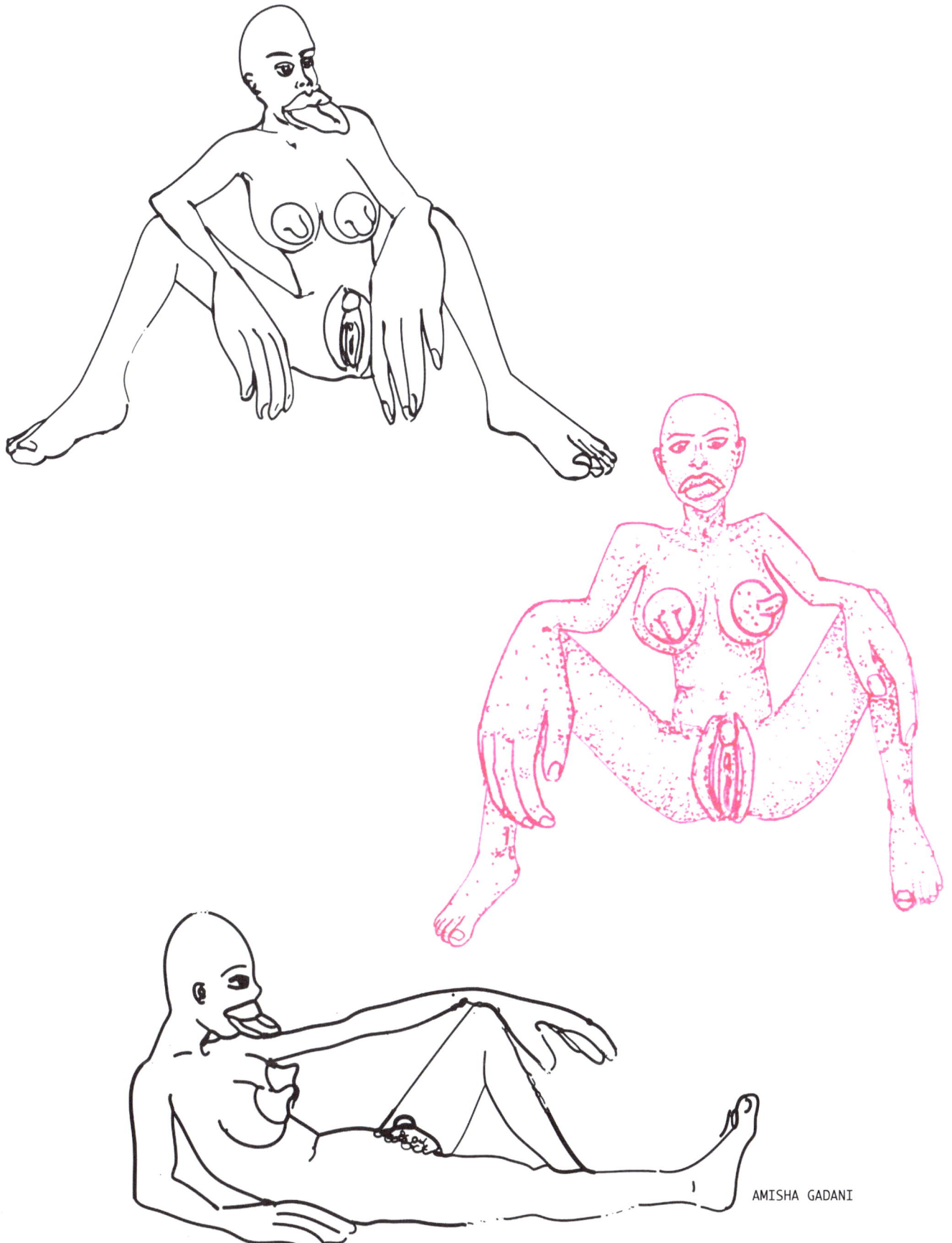

AMISHA GADANI

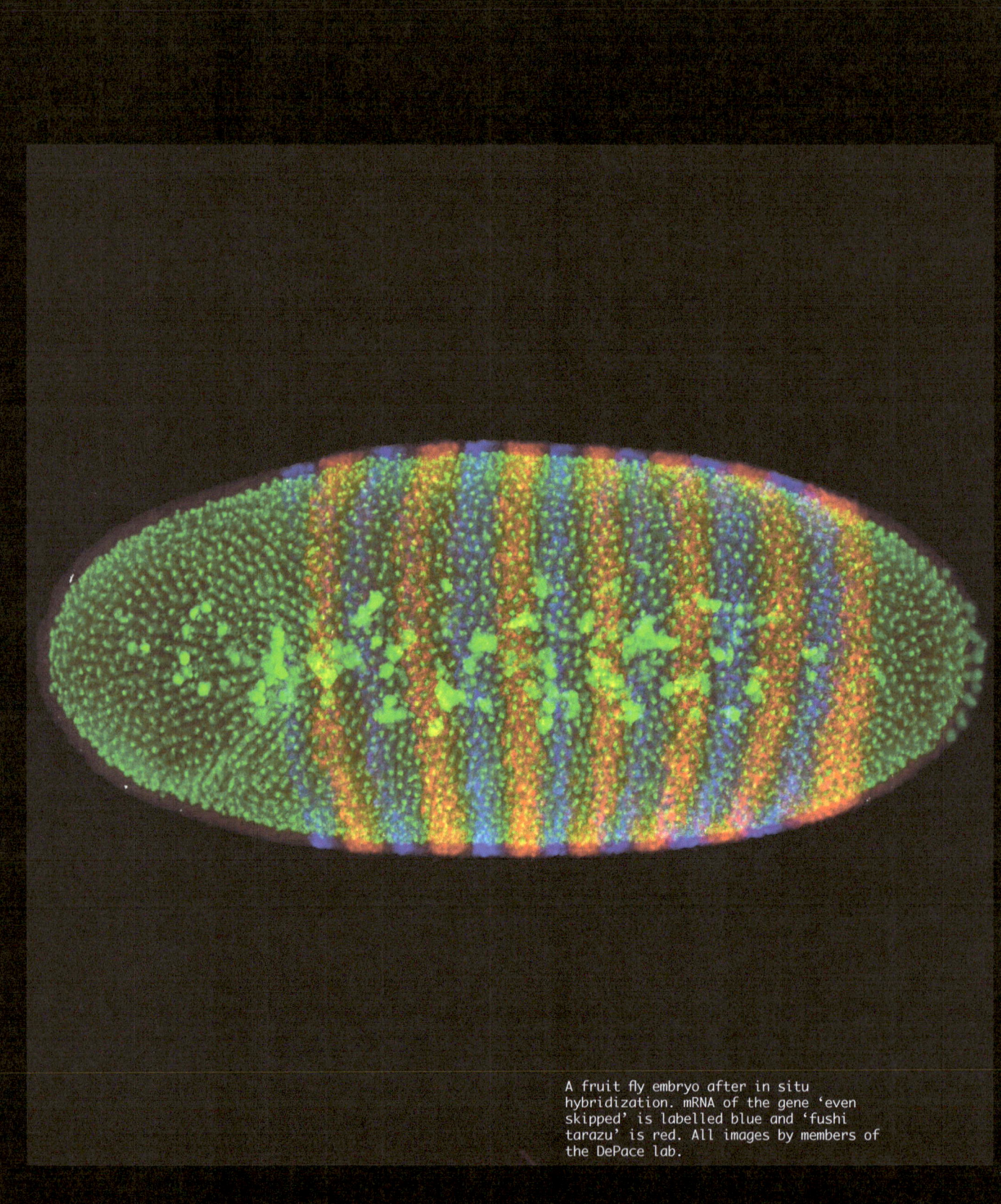

A fruit fly embryo after in situ
hybridization. mRNA of the gene 'even
skipped' is labelled blue and 'fushi
tarazu' is red. All images by members of
the DePace lab.

Ben Vincent, Francheska Lopez-Rivera, Javier Estrada, Jeehae Park,
Meghan Bragdon, and Zeba Wunderlich

Protocol: Visualizing mRNA in Fruit Fly Embryos

Seeing mRNA inside a fruit fly embryo

Animals are made up of many different types of cells, like neurons or muscle cells, which have very different functions. These different cell types are generated during embryonic development, differentiating and continuing to have different function because they make different proteins. However, an animal's cells all have the same DNA, which contains the instructions for makings those proteins, so each cell type must use a different subset of these instructions to make only the proteins that they need. The cell turns on the DNA it needs by first transcribing it into messenger RNA, or mRNA, and then translating that mRNA into protein. To study how different cell types emerge during development through the action of different DNA sequences, we visualize mRNAs in fruit fly embryos.

We study fruit flies because they have a surprising amount in common with humans and are easier and more ethical to work with. Many of the proteins found in a fruit fly are also present in humans, and many of the ways that cell types are generated are similar between these two animals. Therefore, studying mRNA production and development in fruit flies can help us learn about how the process works in humans and why it sometimes goes wrong. Scientists have been studying fruit flies for decades, so there are many experimental tools available for us to visualize mRNA and to genetically modify fruit flies.

A fruit fly develops very quickly. Just three hours after the egg is fertilized, the original single cell has divided 14 times, yielding about 6000 cells. The embryos at this stage look like tiny footballs, with all of their cells on the surface and yolk in the middle. The cells look similar, but they actually already know what part of the body they will become. We know this because embryos lacking specific proteins at this stage develop into larvae with missing body parts. For example, embryos missing the protein produced by the mRNA shown in red below are missing the middle part of their body. In contrast, embryos missing the protein produced by the blue mRNA are missing every other segment of their body. Therefore, knowing where and when mRNAs are produced in an embryo can tell us about their specific function.

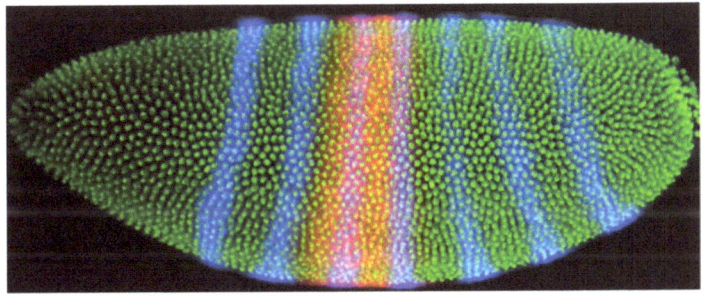

So how do we visualize where the mRNAs are in the embryo? Unlike DNA, mRNA is single stranded, but that single strand can bind very tightly to another single strand with a complementary sequence (A binds with T, C binds with G). We can artificially synthesize this complementary mRNA strand, which we call a probe, and label it for visualization (more on that below). To make a probe, all we need to know is the sequence of the mRNA, which we can get from the original DNA sequence of the gene whose mRNA we want to visualize. The process of seeing where mRNA is expressed in an organism using probes is called *in situ* hybridization. Hybridization refers to the binding of the mRNA and the complementary probe, and *in situ* means that we do the experiment 'in place'—meaning in the embryo. Below is the protocol for using *in situ* hybridization to visualize mRNAs in fruit fly embryos.

1. First, we place approximately 2000 male and female fruit flies in a cage that allows us to collect their embryos. We put an agar-molasses plate on the bottom of the cage, where the flies will deposit their eggs.

2. We change the plates after four hours to ensure that most of the embryos are at the developmental stage that we study. The plates are covered in embryos (each white speck to the right) that we flush with water into a small bucket with a mesh bottom for collection.

3. Once the embryos are collected in the bucket, we can put them in bleach to release the eggshell (yes, fruit flies have eggshells!). We then shake the embryos in a formaldehyde-heptane mixture for 25 minutes to stop the cells from dividing and permanently freeze the embryos at the right age. Vigorously shaking the embryos in a methanol wash will then pop off their outer membrane, resulting in naked embryos. Subsequent methanol and ethanol washes ensure that the embryos are in a pure ethanol environment in which they can be frozen until they're needed for an experiment.

4. Our *in situ* hybridization protocol involves a week of moving tiny amounts of liquid in and out of the tube holding our embryos. This process begins with adding the complementary probes that bind the mRNAs we're interested in and then washing away the unbound excess probes.

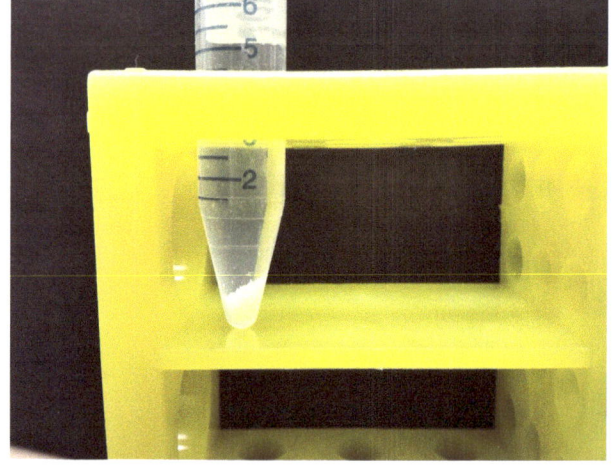

5. The probes aren't visible on their own, so we need to attach a molecule that produces light to them. This is a multi-step process. Initially we synthesize the mRNA probe with an extra chemical attachment at the end. That chemical attachment can bind to a protein that we add after the probe is bound to the mRNAs. That protein can then catalyze a reaction that deposits a fluorescent dye. We can deposit two different fluorescent dyes (red and blue) to visualize two mRNAs at once. After we wash with the protein and then the dye, the dye is stuck in place (*in situ*!) We also stain the embryos with Sytox, a chemical that stains all DNA and fluoresces green, allowing us to see not just the mRNAs, but also each of the 6000 cells that form the embryo.

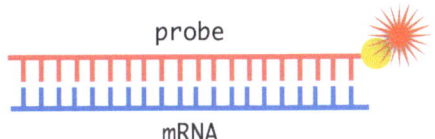

6. After *in situ* hybridization, we are ready to prepare the embryos for mounting on slides. The first step is to clear the embryos with an organic chemical that makes the yolk transparent and allows us to image the embryo from top to bottom. Then we simply spread the embryos onto glass slides and cover them in DEPEX, a viscous liquid that suspends the embryos and protects them from getting squished by the coverslip. We also lay two more coverslips on either side of the patch of embryos to act as legs of a bridge to the overlaying coverslip, further protecting the embryos.

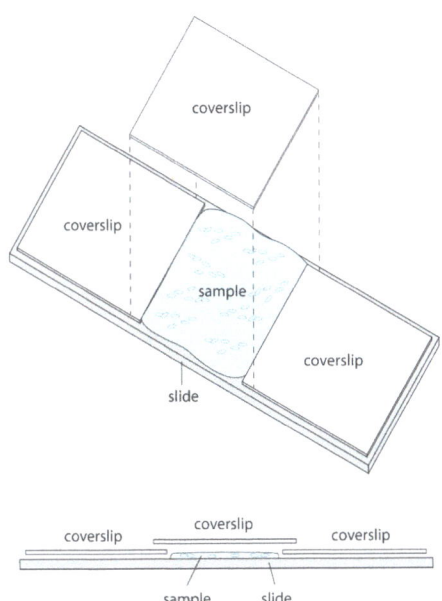

7. The DEPEX dries for 5 days and then the slides are ready to be imaged. Our microscope is a laser scanning microscope, which scans a laser across the embryo to generate a high-resolution image pixel-by-pixel. We need high resolution to distinguish the tiny cells that make up the embryo. The microscope is controlled by a computer with software that records where the embryos are on the slide. We also record the top-most and bottom-most edge of each embryo. After we have found all the embryos on a slide, the software tells the microscope to go back to each embryo and take a 'stack' of 2D images, from top to bottom. These images, when combined, create a 3D image of each embryo.

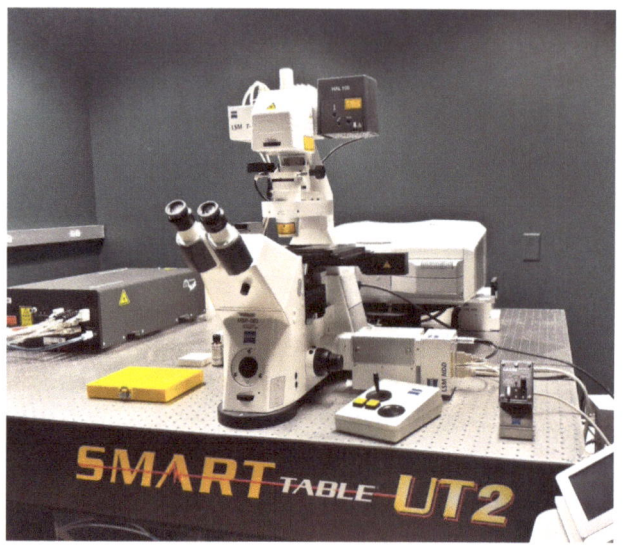

8. We take 3D images of many embryos. For a single mRNA of interest, we take a lot of pictures so we can average the mRNA pattern and reduce the experimental uncertainty of our measurements. Also, there are many different mRNAs that are important for the developing embryo, so we do many *in situ* hybridizations to learn about all of them.

The DePace lab is in the Systems Biology department at Harvard Medical School. The lab studies the mechanism and evolution of transcription in animals, using fruit fly embryos as a model system.

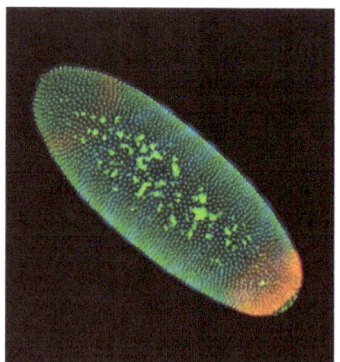 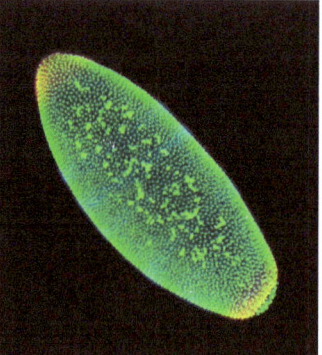 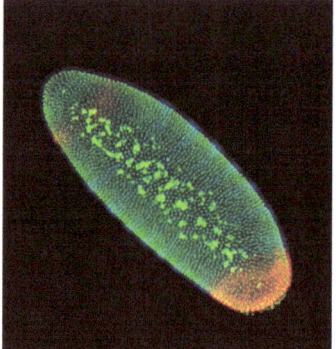 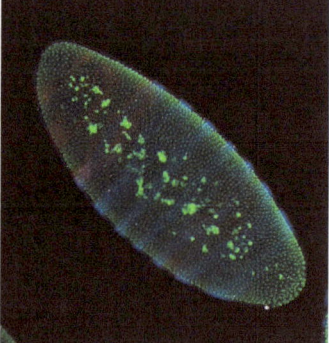

In situ hybridization is used in labs all over the world on many different animals and plants to understand how different cell types are made. In flies, it has revolutionized our understanding of how just a handful of master proteins can lay down the body plan of an animal in just a few short hours. Many of these proteins are used throughout the animal kingdom during development to create a complicated adult animal from a single cell. We owe much of our understanding of human development to the fruit fly and their tiny football-shaped embryos. ■

NASA

Sam Golding

Visions of the Lone Scientific Genius

What the phenomenon of simultaneous discovery can tell us about how we mythologize our lone scientific visionaries in hindsight

The funny thing about science is that it often behaves in ways that seem entirely irrational.

Take the strange phenomenon of simultaneous discovery: two or more thinkers come up with a new idea independently of one another at almost exactly the same time. The invention of calculus by Newton and Leibniz; Darwin and Wallace's theory of evolution; Möbius and Listing's discovery of the Möbius strip. What's so odd about simultaneous discoveries—also known as 'multiples'—is that they aren't isolated events. They happen all the time.

Stranger still are those scientists who appear to have known things long before they could possibly know them. Stories of such people are incredibly rare and must be treated with caution. Like that popular misconception, the Eureka Myth, narratives of visionary foreknowledge tend to be apocryphal, presenting discovery as a result of individual insight rather than a process of investigation, calibration, and collaboration. But among the noise, do real visionaries—those who appear to 'see' the scientific future—really exist?

Simultaneous discovery and visionary foreknowledge hold a special place in the imagination because they appear to defy logic and reason. But humans, not machines, conduct science. And as such, discovery can be organic and surprising, unpredictable and messy. Discoveries are made, not found. If these stories really are true, how and why they happen—and whether they constitute an intrinsic part of scientific research or are a condition of temporally specific contexts—are important questions.

But they also make for fascinating stories.

■ ■ ■

For a while, no one noticed anything different. Palace Terrace Gardens was just an ordinary square in the heart of leafy west London. But when a new tenant moved into Number 8, things started getting strange.

This man was odd. He would pace the street into the early hours of the morning, muttering under his breath accompanied only by his dog. Some said the dog was more akin to a lab rat, that it was the subject of bizarre electrical experiments—its fur frayed from static charge. And then there was the coffin. Eight feet long and painted black, propped up by the attic window. For hours, the man would gaze into it, stopping only to make frantic notes. Occasionally his wife would join him; other times they would invite their friends to have a look too. What was in there? Was the man conducting some kind of

JAMES CLERK MAXWELL IN POPULAR SCIENCE, 1880

freakish experiment, trying to reanimate dead matter?

The coffin in fact turned out to be a light box: a homemade device capable of producing an array of spectral colors. And the man was James Clerk Maxwell, the pioneer of electromagnetism. The above quirks are all true, but it goes to show how easy it is to paint a caricature of the lone, eccentric scientist using a few scraps of biographical information.

And yet. It seems as if Maxwell really was a brilliant visionary 'ahead of his time.' He discovered chaos theory. And he did so 100 years before it was even recognized as a science.

Maxwell's path to chaos began when he turned his mind to one of the most mysterious objects in the cosmos—Saturn's rings. For over two centuries astronomers had been baffled by their very existence. Were the discs solid or liquid? What were they made of? Why didn't they break apart?

Applying existing mathematics in entirely new arrangements, Maxwell made some startling observations. He realized that a solid ring couldn't be stable. Even the slightest perturbation would create an amplifying vibration that would destroy the entire structure. According to Maxwell, the rings, in spite of their continuous appearance, were in fact comprised of millions of smaller bodies orbiting the planet. For his efforts, Maxwell was awarded the prestigious Adams' Prize for mathematics. But his work on Saturn's rings had also got him thinking about statistically modeling dynamic systems.

In 1858, Maxwell read a paper by Rudolph Clausius that argued that gases are comprised of countless molecules moving around at immense speeds. The kinetic theory of gases could account for a vast range of observed phenomena. But for the theory to explain pressure, Clausius had to assume that the molecules in a gas of a given temperature all moved at the same speed. Maxwell thought that this could not be the case. He was convinced that the speed of individual molecules was random and defied the scientist's powers of observation. The following year, Maxwell presented his work on the topic. His intervention was to devise a statistical function to calculate the overall distribution of molecular velocities and present them as an average. Viewed statistically, a gas's behavior could thus be modeled as a probabilistic function.

This was a profound conceptual shift. It allowed scientists to describe a system of which their knowledge was incomplete. It also heralded the beginning of a new branch of physics, statistical mechanics.

Maxwell's work also contained a radical implication about the relationship between mathematical laws and physical reality. It suggested that the second law of thermodynamics—the law that tells us that over time, the amount of useful energy in a closed system decreases—is not absolute. It is only highly likely. In theory, this means there is nothing to stop a gas's molecules from becoming more ordered, say by returning to a perfume bottle after diffusing throughout a room. Taken even further, it suggests that heat can flow from a cold to hot body. Both cases represent a net decrease in entropy: a fundamental violation of thermodynamics. Maxwell would later expand on this idea in his infamous thought experiment known as 'Maxwell's Demon'. But at first,

James Clerk Maxwell's lightbox. Image: Cavendish Laboratory.

he didn't see this implication. It seemed to emerge autonomously from the fabric of his mathematics.

These ideas led Maxwell to the fundamental tenets of modern chaos theory. He saw that hidden order underlies chaotic systems and that self-organization, or local decreases in entropy, emerges spontaneously from chaos. Moreover, his work on Saturn's rings prompted him to think about sensitive dependence upon initial conditions, popularly known as the 'butterfly effect.' By 1873, Maxwell had a profound understanding of chaotic systems:

"When the state of things is such that an infinitely small variation of the present state will alter only by an infinitely small quantity the state at some future time, the condition of the system, whether at rest or in motion, is said to be stable; but when an infinitely small variation in the present state may bring about a finite difference in the state of the system in a finite time, the condition of the system is said to be unstable."

"I have created a new universe from nothing."

Decades before anyone else it seems, Maxwell understood chaos. Yet were his ideas truly 'visionary,' or part of a more complex framework of discovery? And what does it mean when visions strike more than one 'lone genius' simultaneously?

In the early 1820s, a Hungarian mathematician called Janos Bolyai came up with an idea that would revolutionize geometry. At the same time, 1500 miles away in the Russian city of Kazan, Nikolai Lobachevsky would make the same discovery. And then a third man, Karl Friedrich Gauss—dubbed the 'Prince of Mathematics'—would claim that he too had also come to the same conclusion. And not only that: he had been thinking about this idea for years prior to both Bolyai and Lobachevsky.

The idea stemmed from a challenge to the laws of Euclidean geometry, the basic laws of shape we all learn in school. The internal angles of a triangle add up to 180 degrees; space is isomorphic and flat. It's the mathematics that relates to our subjective experience of space and which allow it to be mapped in three equal dimensions—x, y and z.

Bolyai and Lobachevsky began to challenge the supposed inviolability of these assumptions: in particular, Euclid's parallel postulate. This says that two straight lines, parallel to each other and extending indefinitely, will never meet. Bolyai and Lobachevsky disagreed. They began to re-evaluate the notion of 'straightness' and realized that it only made sense in flat, even space.

They imagined straight lines not as ruler-edges on flat surfaces but rather as lines on a curved surface, such as a saddle shape. This geometry, termed 'hyperbolic,' deals with spaces of constant negative curvature—where every plane through a point is drawn inwards. Lines traced on such spaces violate Euclid's ancient maxim. A hyperbolic triangle's internal angles total less than 180 degrees, while parallel lines diverge away from each other.

The corollary of this is also true: if two vertical lines are traced on the Earth's surface, both beginning at right angles to the equator, they will eventually meet at the same point, the North Pole. These lines, fixed to a spherical, two-dimensional space, are 'straight.' Yet the shape they trace—a geodesic triangle—contains three interior right angles adding up to 270 degrees.

The invention and discovery of so-called 'non-Euclidean' geometry was a profound conceptual shift. Without it, Einstein's relativity wouldn't exist. Bolyai was certain of the uniqueness of his discovery: "I have created a new universe from nothing," he wrote to his father in 1823. Similarly, in an 1824 letter to Franz Taurinus, Gauss claimed that he had found "a curious geometry, quite different from ours, but thoroughly consistent." Lobachevsky, meanwhile, first reported his findings to the University in Kazan in February 1826, but wasn't published until a few years later.

All three men came up with the same idea, at the same time, totally independently. All thought they were alone. And this all happened well before the modern era of global

WIKIMOL

connectivity and interdisciplinary research. Was the time just right? Was something 'in the air'?

Well, in a way, yes. In 1960, a study by Robert Merton claimed, "the pattern of independent multiple discoveries in science is in principle the dominant pattern, rather than a subsidiary one." Subsequent research has come to the same conclusion. In a survey of hundreds of new patent applications, Mark Lemley found that "almost all of them [were] invented simultaneously or nearly simultaneously… independently of each other."

Comparable studies invariably affirm the same thing: simultaneous discovery isn't the exception; it's the norm. In popular accounts of scientific history, scientists are still generally painted as the romantic outsider whose visionary brilliance sparks new ideas ex nihilo. But in reality it doesn't work this way. Scientists, as Lemley writes, "build on the work of those who came before, and new ideas are often 'in the air,' or result from changes in market demand or the availability of new or cheaper starting materials."

Synchronicity in science suddenly seems less surprising. Given that new ideas are so reliant on technology and the ability to connect emergent and disparate areas of knowledge, it's almost inevitable that discoveries will occur multiply. What's more, it's a trend that has been increasing over time.

But what about cases such as Maxwell's, where new ideas are so distinct from their surrounding intellectual context as to appear nonsensical? In these cases, future science seems quite literally made, not found.

Some argue that, unlike the prevalence of simultaneous discovery, individual insight is a phenomenon consigned to science's past. A 2007 study found that a defining trend of contemporary research has been the overall shift from knowledge produced by individuals to teams. Analyzing nearly 20 million papers from the last five decades, the researchers found that teams produced both the vast majority of "exceptionally high impact research" and work dominating citation distribution. This is perhaps due to greater recognition of the role lab assistants and technicians play in gathering data. But it's also a phenomenon driven by the increased subdivision of disciplines.

Northwestern professor Ben Jones suggests as much in a paper looking at the changing conditions of innovation. Jones claims that over time, flashes of scientific insight have declined—a result of the multidisciplinary nature of experimental work and the increasing specialization of knowledge. "If one is to stand on the shoulders of giants," he writes, "one must first climb up their backs, and the greater the body of knowledge, the harder this climb becomes."

Our skepticism towards narratives of visionary genius also stems from a greater awareness of how science is reported. Histories are constructed accounts that tend to mythologize the individual in place of the messy and evolving reality of scientific plurality. This is not to take away from Maxwell's extraordinary intuition, nor is it to suggest science is practiced the same way throughout history. 19th century Britain certainly had fewer professional scientists than today, working in broader knowledge bases with a lower overall body of specialized information. If an individual was going to make a revolutionary discovery, then was as good a time as any.

But Maxwell also didn't work in a vacuum. He processed information analogically, taking ideas from disparate branches of mathematics and applying them in new ways. He conducted experiments with other scientists, went to conferences, conferred about speculative ideas. And his own models contained implications of which he wasn't always aware. Implications that bubbled under the surface but weren't fully recognized until the fog of intellectual confusion had time to dissipate.

The most surprising thing then about Maxwell's understanding of chaos is that it doesn't fit the standard model of simultaneous discovery. He just happened to connect the dots in the right way. He was able to do so because he was a brilliant scientist, but there were numerous advantageous variables at play as well.

As such, stories of visionary science are problematic because they take a singular view of multiplicity. Maxwell's case was a statistical anomaly, and one that itself can be considered multiple given his reliance on other's work. But statistics can help us to better understand 'insight' and simultaneity and refine the way we construct historical narratives. Because, to use a Maxwellian analogy, scientific practices are themselves dynamical systems: constantly in flux and comprised of countless variables. We can't trace these variables individually. But we can map them statistically by writing stories that more accurately approximate the unstable, fraught conditions of discovery. ■

Sam Golding is writing his doctoral thesis on the intersection between Victorian science and culture at King's College London and works as a journalist for the Associated Press.

Christina Agapakis

Conversations with Natalie Jeremijenko

Natalie Jeremijenko is an artist and engineer whose work explores the intersection of human and environmental health. In a series of conversations, we spoke about data, visualization, and the process of making sense of complex phenomena. The following is a heavily edited transcript of her words.

Making Sense: Gravity Probe B

If we're alert to our own ways of making sense of context and what really works to make sense and what we trust, we can start to really question how we visualize and how we make sense of the world.

In my engineering work, working in precision engineering on Gravity Probe B at Stanford with some wonderful people, I learned a lot about physics but also began to really understand how I was making sense of things. Gravity Probe B was the only academic satellite ever launched. It was to test the warp of space-time, testing general relativity theory. While I love physics and the opportunity to dive into gravitational theory was fantastic, what was really sense-making for me—in terms of how to make sense—was building this precision optics system, which was so much about a fairly humble material understanding of the quartz ball that was in the gyroscope that measured the warp of spacetime that was held by this precision optic system inside the satellite that I was working on. That this was the roundest object ever made in the history of the world.

To me, this idea of translating, in a very material object and through very material practices, these complex physical phenomena is very powerful. Being able to twist a ball on a bed sheet to intuitively test your understanding of how spacetime worked, to be able to think through a hard mathematically-described phenomenon with these radically different forms of representation is of course how we make sense of things. It's not being able to perfectly understand it in one language, it's about being able to make those metaphorical contrasts. You can prototype a sort of understanding. Think about torque—you can quote the physics slogan of what torque is but it's actually physically bending and twisting your pen that can really make concrete what the forces involved mean.

NASA

Christina Agapakis is a biologist and writer. She is a founding editor of Method Quarterly.

I think hands-on, problem driven learning, and the shifts in engineering education towards this kind of material practice is about helping people to think across these domains, because doing lots of problem sets doesn't make you a great engineer. It doesn't develop your intuitions. So to some extent, there's a recognition that hands-on, peer-driven, socially-situated, contextualized problems make sense.

This is related to research I did on the attrition of women and minorities form engineering fields. This was in the early 90s, when the NSF had these huge consortiums of engineering schools and they had six or seven major engineering schools looking at this issue, doing exit interviews with women and minorities that had dropped out of engineering to see what they said and justify the fact that in general and all along the pipeline, women and minorities tend to get higher GPAs, yet drop out at a higher rate. The traditional way to think about this is to sort of talk about the hostility towards women and minorities in these engineering and technical fields, or at worse to say something like "women are just not good at it." Well, that's just not the case!

What I came to understand it as, is not as a problem of women and minorities somehow not being able to acculturate, but as a massive social protest. If you looked at all the exit interviews, they all basically said, "I want to help people, I want to do something that helps people," — this naïve way of saying that the militarized, DARPA-funded and corporate engineering being taught is just not interesting, it's just not what young people want to do. So the attrition rate is more a massive social protest that this is not the kind of work we want to do. [It says,] we want to do something that's more meaningful, taking on a huge and important challenge, instead of building a better ballistics system, or a robotic arm for industrial assembly that has an extra degree of freedom or tactile manipulation — projects that just are not that compelling.

Playful Data: Feral Robotic Dogs

So to transform engineering education, you create problems and issues that are socially embedded in real compelling political contexts, and the feral robotic dogs project was an example of that. We start with these consumer toy robots. You open them up and you can see what was hand-soldered, you can see the labor that's gone into them and the ingenuity, you can gently amputate the legs. You're critiquing a manufactured object, in some sense reverse-engineering it to make it perform in an all-terrain context — what the feral robots league was really about. And of course you're updating its *rasion d'etre* by adding the environmental toxin sensor, and by adding an additional microprocessor. So you program the dogs to follow the concentration gradient of the environmental contaminant that they were being exposed to.

The sensors we put in the dog's noses were measuring VOC — volatile organic compounds — the most ubiquitous urban pollutant. Then we designing the dogs to release them on public contaminated sites, and with their movement they display data about pollution in the area. We were capturing data that we could later analyze, but the immediate sense-making was, "Oh, the dog is going over there! Oh, the dog's chasing a car!" They actually kept going towards the nearest road, which of course ends up being the biggest lesson of these robotic dog packs that we released all over the country was that what dominates is not the subsoil contaminants, but the traffic — the cars.

But what would then happen is that journalists and other people would turn up and they would ask the students who worked on the dogs, the dog trainers, "What's your dog finding? What does it mean? What do we do about it?" And of course the students had been working on their dogs and thinking about this site and they can answer, they address it. It's not about me collecting all this data inexpensively with these low-cost robots and student labor, it's actually the collective issue of making sense of a very tricky problem. There is no absolute answer about what the contaminants are in the environment and how they're affecting human health. There are these contaminants, and the data helps us to make sense of it, but it's also really a collective process of asking people to think about it.

Through these conversations we might start to make sense of what could be done, what should be done? What do you do about VOCs? Is this an issue that's totally dominated by nearby road traffic or are these hot spots? How does this

MEDDYGARNET

Phenological clocks for Sydney (top) and New York (bottom) by Tega Brain.

contribute to asthma rates in the neighborhood? So all these questions are not ones in which any expert has the definitive answer, but what's important is facilitating this sort of project where we can make sense of it without pretending that the data is going to tell us everything—that meaning is going to drop out of the data.

Meaning and sense-making are social processes. In a sense, all of the projects I've done are in this realm of trying to make sense of these irreducibly complex socioecological systems, in which there isn't *a* simple experiment or *a* paper, or even *a* discipline or *an* expertise that can alone figure out what we can do. These are questions that can lead to the crisis of agency that we all face, what do *I* do? What does little old me do? What does any one of us do in the face of so many challenges?

Beyond Snapshots: The Phenological Clock

I'm very interested in using the authority of natural systems to reinvigorate the imaginative capacities of political and social structures of participation. So I'm very specifically interested in finding mutualistic relationships to amplify, letting them emerge from the data, and correcting for the systematic blindness to mutualistic systems.

With the phenological clock, the effort of this visualization is to reveal the temporal interdependence. There's a lot of things we've been looking at, to look at the shifts over the years, and how the pollinator that depends on the flower, if they shift out of sync, then with the birds that depend on the insect, you'll see that. I think this works as an icon, just to say, it's not a field guide, these organisms are always in relation to each other and that it's a dynamic, temporal system that's not a photograph. Which I think is the opportunity that data presents us, rather than doing that slice in time.

Compared to the snapshot idea—that you could take a snapshot here, maybe a snapshot there—the phenological clock privileges that these systems are always and already changing. It's about these seasonal phenomena that are very material and specific and constituted by the complex ecosystems that occur.

This is still very much in development, but I'm really interested in how it invites further observation, that in itself it's not visualizing the data retrospectively, it's more that it invites having it around and ubiquitous in a public platform, on a watch, or in this kind of non-science venue. That it would invite people to think, "Oh, it's daffodil time," or, "The tomatoes are starting to grow," that it becomes interwoven with social, lived reality. Where I went to school, the jacarandas always bloomed when it was exam time, so there was always a kind of dread when we saw these beautiful purple flowers. The idea is to start to weave together social events and reconstitute convivial social lived experience with these natural phenomena.

It also invites the Instagram image where you can tag it with the project hashtag and it has the species, and it has the geolocation, and it has the time—which gives us all the information we need for the clock. It's already a populist thing to do, and yet it underscores that simple kind of wonderment and the act of observing. It's underscoring our collective attention. That your attention is contributing to our collective understanding.

To some extent I don't like the idea of data visualization, because there isn't that citation or the reference there to be able to question their assumptions or their method. You don't even have error bars, or any of the cues that—if you're familiar with the dataset or familiar with the argument that you would use—it obfuscates. It does the reverse of inviting participation, it closes it, says, "It's done, we thought about this, it's done, it's over, you don't need to think about it."

This idea of data spectatorship—this passive consumption of data visualization—that's what I'm working against, the data spectator. I suppose it's about going from data spectator to data contact sports or something.

Natural Intelligence: One Trees and the Mussel Choir

With the One Trees project, we planted 80 genetically identical trees in the same place, in pairs, so you could see the within-pair and between-pair differences, and visited them over the years to try to make sense of why do these trees look different? Why have they diverged so extraordinarily?

So I was looking at this pair of trees at 22nd and Valencia in San Francisco and there was a man working construction across the road. He saw me looking at these trees and he came over to ask me what I was doing. I told him about the project and how these trees are about 12 years old and I can't make sense of why they look so different. I thought one had gotten into the water mains and the other hadn't, that it had gotten a taproot in. But I found they actually use terra cotta mains in San Francisco, which means that since it's a seismically active area, all those terra cotta mains are cracked and leaking and subsidizing the entire urban forest. So it's not that this is hermetically sealed system.

I just could not make sense of it. I looked at leaf width, looked at solar exposure, looked at everything I could to figure it out and I couldn't figure out! And then, after I explained the project to him, this guy working in construction said, "Oh it's totally obvious why they're different!" and I said, "It is?" He said, "Well, look behind the trees." One has a little 1950s one-story building behind it, the big one has an old Victorian behind it, two stories and lovely. He said, "Well between those structures was the 1906 earthquake." And then he had to explain to me that that meant that building codes had changed and foundations had radically changed and probably that what I was seeing was that the tree in front of the small structure is like a massive bonsai—it has a strong foundation that blocks off the roots—whereas the one under in front of the Victorian probably had roots that extend all the way through the and under the Victorian. This was the best explanation I'd had. I wanted to tell that story because that's coming not from an arborist's point of view, or soil science, or stomatal blockages from particulate matter, or ozone, or any of the issues that we thought.

Another time there was a group of students from the creative writing studio from around the corner that came by and one of the kids wanted to watch for which tree the birds preferred. I was so struck by this idea, that perhaps somehow the birds could make sense of this better than we could, which I thought was another delightful idea about how we could observe and make sense of urban ecosystems.

One Trees at 22nd
and Valencia in San
Francisco. Photograph by
Marina Agapakis.

THE MUSSEL CHOIR. IMAGE: NATALIE JEREMIJENKO

The Mussel Choir project is also about looking to animals in order to give us this kind of information about complex ecosystems. The first hit hissingly of the mussel choir is "The Bicycle Built for Two," that goes, "Daisy, daisy tell me your answer true." This is an icon of artificial intelligence because HAL sings it in 2001, Siri sings it, it was the first computer-synthesized voice—so it has become this icon. Aaron Koblin did this lovely piece on distributed intelligence, using 2000 Amazon Mechanical Turkers, having them each sing "aah" at different notes and putting them all together into "A bicycle built for 2000." The hit single for the mussels is "A Bicycle Built for Too Many," which is sung by the mussels. As they open and close, I'm sensing their gape angle and converting that to synthesized sound.

The mussels are responding to the local conditions, integrating over the many different parameters, many more than we can measure in our battery of water quality tests. They are of course responding intelligently, as if their lives depended on it! I use that to have them sing "The Bicycle Built for Too Many" in order to iconify and contrast to the AI approach, the artificial intelligence approach, or this distributed intelligence approach. I call this the natural intelligence approach, NI — as opposed to AI — which is that we can draw on the sense-making of mussels or the birds choosing one tree over the other, or how the trees themselves respond to their environment.

This kind of making sense of the material world with these radically different representations that come from diverse organisms is critical for making sense of these complex urban ecosystems. They give us this immediate feedback cycle, where we're not just looking at some sort of blank array of distributed low power sensors. The vision of the "smart city" is that we'll have these sensors everywhere, electronic sensors that will somehow make sense of things. But we can see that with the battery of what we use to make sense of water quality — dissolved oxygen, turbidity, pH, salinity — even water quality experts can't tell you what that means. But if the mussels can survive, if they can thrive, or if they don't thrive, or if their flapping in panic, or if they're clammed shut and not singing—I trust a mussel more than the data! There can't be any decimal point errors or recalibration issues when their lives depend on it.

So this idea that we can make sense through and with these diverse representations and these non-human organisms and is really part of the story of how you make sense of things in order to act on them, to improve and change our relationship to natural systems. The goal is to redesign this relationship so that we actually can increase biodiversity and increase water quality, using living infrastructure to make these places more habitable. ■

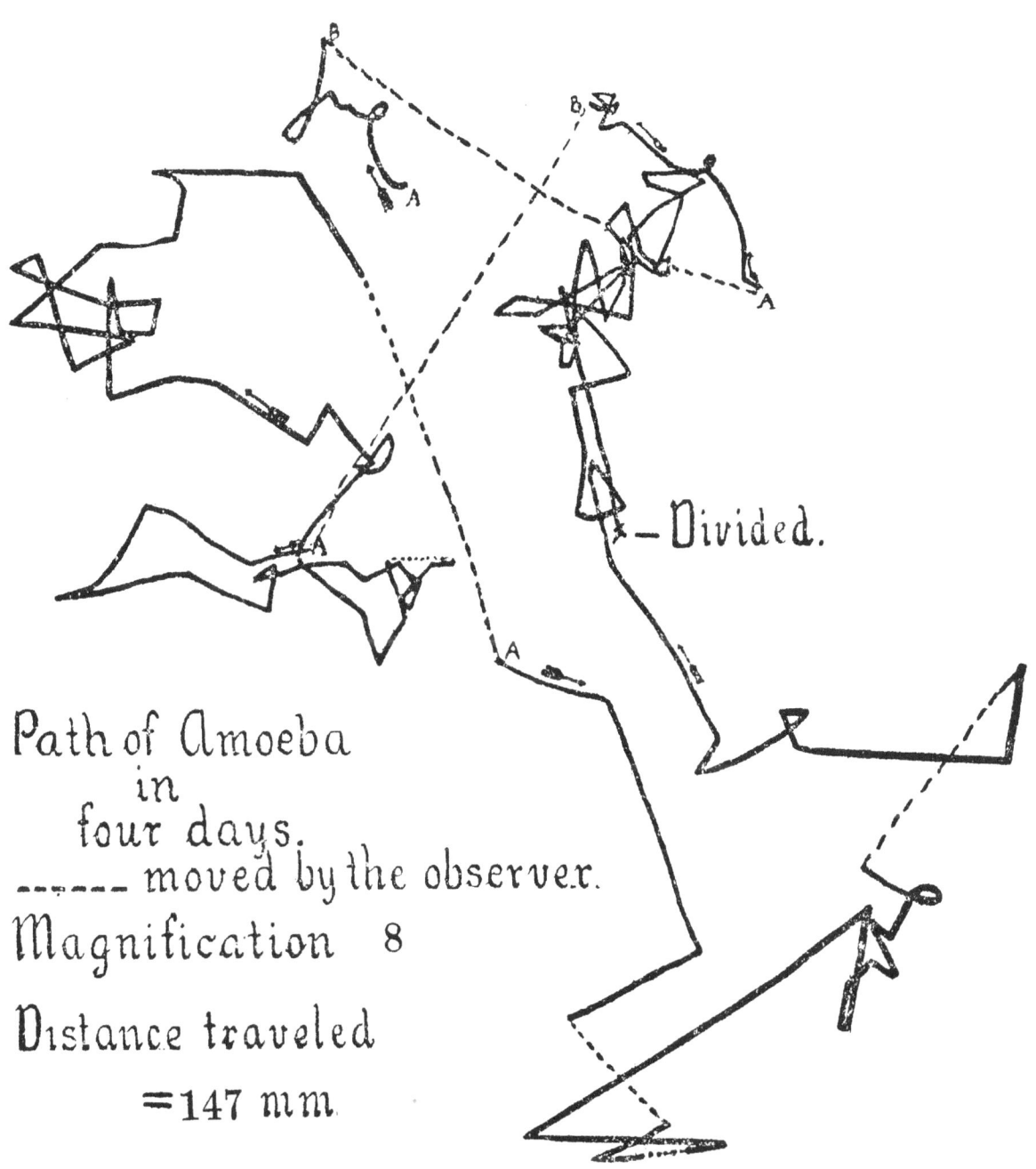

Path of Amoeba
in
four days.
------ moved by the observer.
Magnification 8
Distance traveled
=147 mm.

—Divided.

FIG. 3.

Azeen Ghorayshi

The Daily Life of Amoeba Proteus

Seeing amoeba and seeing ourselves

In late 1905, two biologists from Clark University—Oris Polk Dellinger and David Gibbs—worked in a continuous relay for six days and five nights to observe several *Amoeba proteus* in their "daily lives." Recording their hourly movements, Dellinger and Gibbs tried to discern whether the amoeba's life patterns fell into the rhythmic ebbs and flows of work and rest—tied to the search and attainment of food—that were the basis for all known animal life. Observed under a microscope, diagrammed in their zig-zagging movements, and photographed landing their prey, the biologists found that the amoeba too exhibited these characteristic patterns. Viewed in the context of their behaviors, Dellinger and Gibbs concluded, the amoeba must be considered a rightful member of the animal kingdom. Much anthropomorphizing ensued. The following quotes and images can serve as a summary of their 1908 paper, "The Daily Life of *Amoeba Proteus.*"

■■■

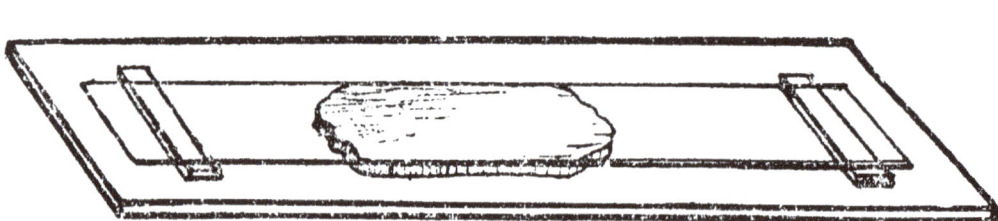

FIG I.

INTRO: AMOEBAS ARE EXISTENTIALLY IMPORTANT

"The question of the behavior of amoeba is not a new one. The movements of no one animal have been studied so repeatedly, so carefully, for so many years, and so frequently referred to as those of the amoeba…The amoeba figures in discussions of immortality, heredity, and death."

STEP 1: PURSUIT

"The amoeba often follows a paramoecium or a ciliate until it is caught or lost. This 'pursuit' may continue for twenty minutes or more as is indicated by the drawings of an actual instance, Fig. 7. When in 'pursuit' in this way the amoeba does not generally respond to other stimuli, especially if close upon its prey."

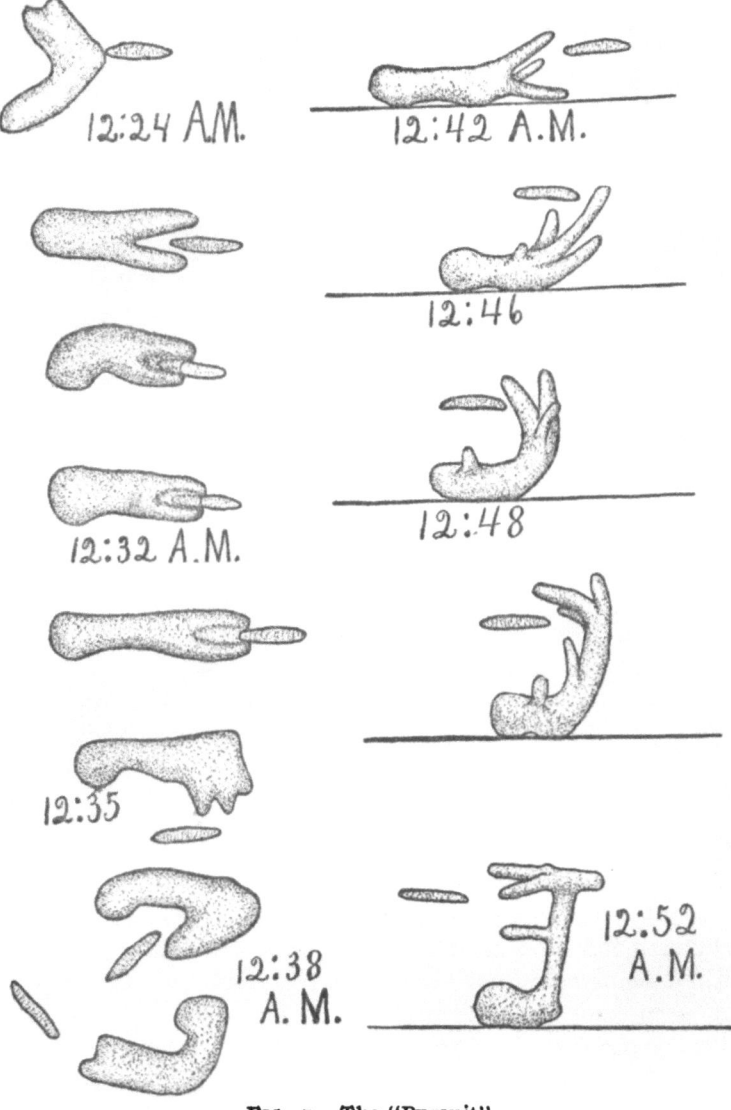

FIG. 7. The "Pursuit".

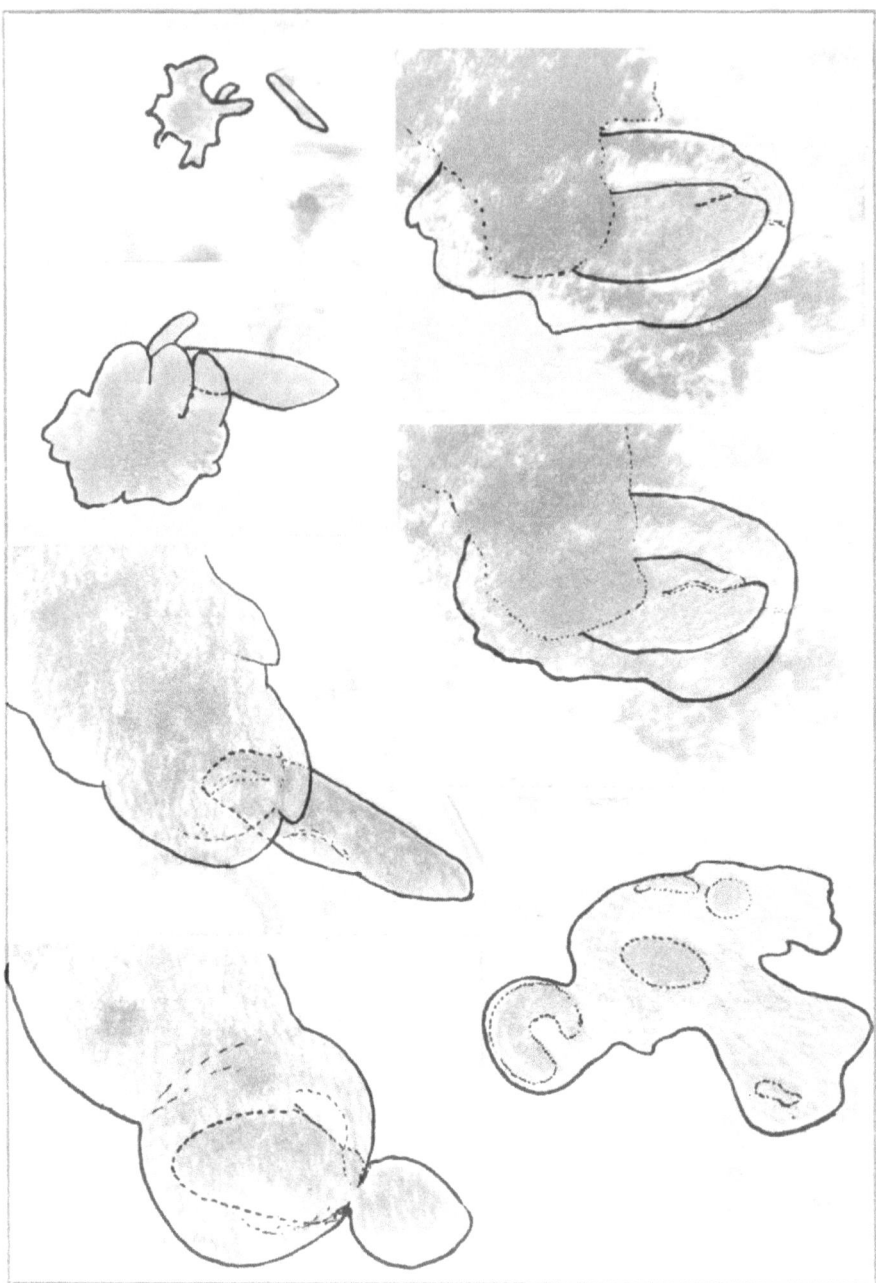

STEP 2: CONFUSION

"An amoeba suddenly placed in the midst of a large number of paramoecia, which bump it and knock it about, usually makes no response to the separate stimuli, but seems 'confused.' Later, some amoebas in these circumstances, put out pseudopods and may 'pursue' a single paramoecium without much regard to touches from the others; while some appear never to get their equilibrium, but move off or take the spherical form."

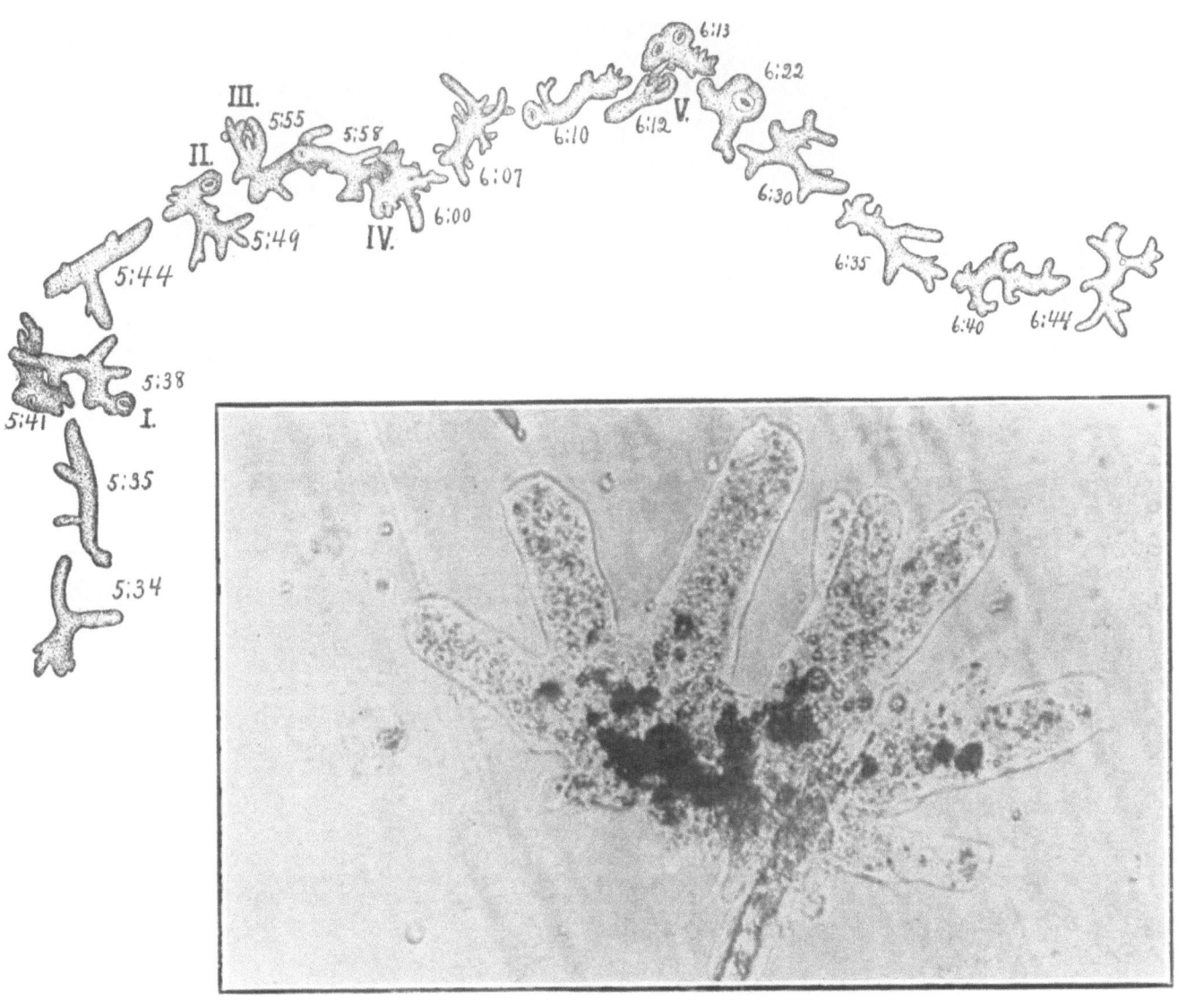

PLATE I. Au amœba feeding on algæ.

STEP 3: CAPTURE + FEEDING

"The intensely interesting sight of an amoeba after numerous trials gradually sliding its pseudopods around a feeding paramoecium, throwing a cover over it, closing the pseudopods, and gradually squeezing the struggling victim down to a rounded mass can hardly be described without using anthropomorphic terms. It requires a number of adaptations and considerable skill, which our observations seem to indicate are acquired by the 'method of trial and error.'"

STEP 4: REST...BEFORE STARTING ALL OVER AGAIN

"Activity, feeding, rest; activity, feeding, rest...Activity, or the performing of work, requires energy which the protoplasm must supply. The period of rest appears to be simply the result of organic satisfaction, or a period of recuperation. It suggests the lowest form of sleep; for this tendency to rest, to sleep, as a food reaction is illustrated by the higher animals. In this respect this lowest form of life does not differ essentially from the higher forms."

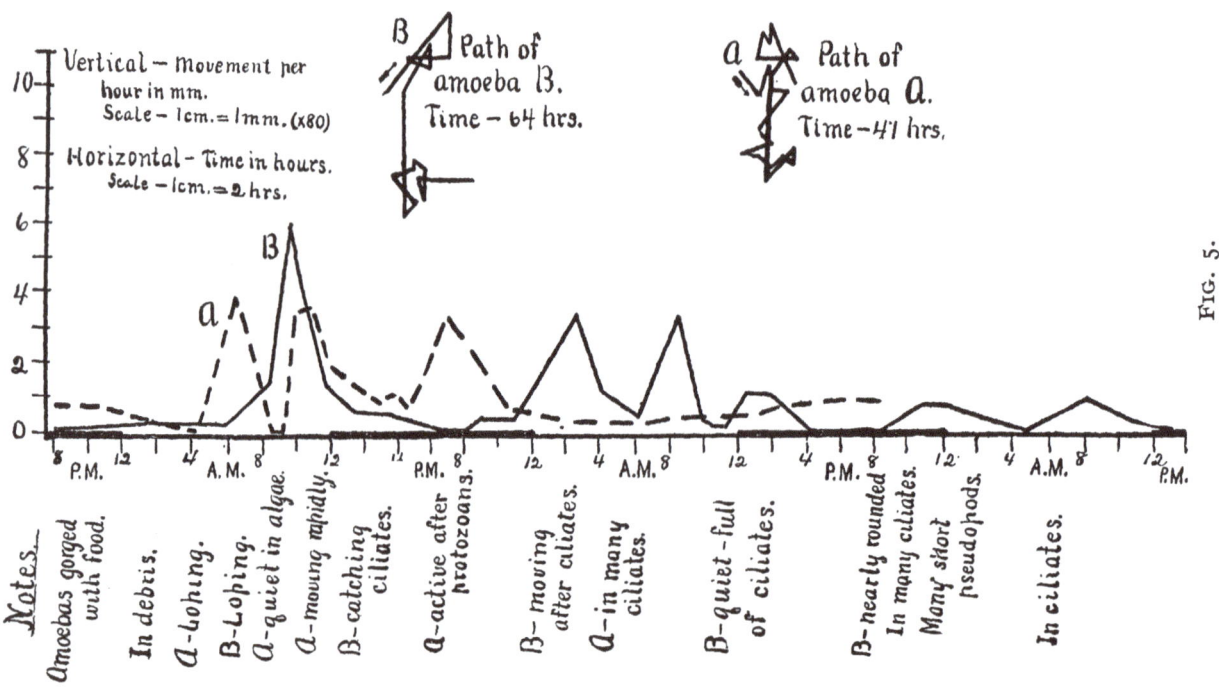

Fig. 5.

CONCLUSION: AMOEBA LIVES MATTER

"The study seems to show that amoeba can no longer be considered as a bit of but slightly differentiated protoplasm, but must take its place in the true animal series with the rudiments at least of true animal behavior." ■

Azeen Ghorayshi is a science writer and a founding editor of Method Quarterly.

*Thanks to UCSF bioinformatics graduate student Patrick Harrigan for telling Method about this paper.

Zuri Sullivan

Seeing the Future of African Science

The ingredients—tangible and far less so—needed to build a world-class research institution in South Africa from scratch

Between 2005 and 2006, an outbreak of extensively drug resistant tuberculosis (XDR-TB) killed all but one patient at the Church of Scotland Hospital in Tugela Ferry, South Africa. The median survival time following diagnosis was a mere 16 days, and of the patients tested, all were co-infected with HIV. The situation was desperate, the fatality rates unprecedented, and the community unprepared for an outbreak of this magnitude. Though this calamity sent shockwaves through the TB and HIV research communities, the situation was not unique. XDR-TB had been detected in all of South Africa's nine provinces, all of its neighboring countries, and dozens of other countries across the globe. HIV was fueling the TB epidemic, and Africa was the

only region of the world in which TB incidence was on the rise. No new TB drugs had been discovered in nearly 40 years. Effective vaccines against HIV or TB infection remained a dream.

More than 8,000 miles away, in Chevy Chase, Maryland, American researchers at the Howard Hughes Medical Institute (HHMI), the largest private funder of academic biomedical research in the United States, were meeting to discuss their international program. At this meeting, Dr. Bruce Walker, an HHMI investigator who leads an HIV research lab at Harvard Medical School, proposed using a model similar to what has historically worked for HHMI in the United States: investing in individual investigators who were doing great work in biomedical science.

"But," Dr. Walker recalls, "In a place like Africa, truly transformative support would require establishment of critical infrastructure and a critical mass of investigators." Indeed, this meeting identified two important and intricately related problems in the developing world: the desperate need for cutting edge biomedical research, and the rarity of sites in which to conduct such research. These issues were not new to Dr. Walker, whose longstanding collaboration with the Doris Duke Medical Research Institute (DDMRI) in Durban, South Africa led to his suggestion that this model could be expanded, "and ultimately used to repatriate African scientists who wanted to work in Africa but were lacking opportunities."

The plan evolved as Dr. Walker traveled to Durban to meet with leading South Africa-based scientists in HIV and TB research, building a committee that worked over the next three years to realize their vision for the KwaZulu-Natal Research Institute for Tuberculosis and HIV (K-RITH). Back across the Atlantic, at Harvard College, I was working to realize my much smaller vision of graduating and pursuing international infectious disease research.

As a rising senior in 2011, I was starting to worry about where I would be after graduation. I had spent three years working in a basic science research lab studying *Mycobacterium tuberculosis*, the bacteria that causes TB. Scientifically, I was interested in the interaction between HIV, TB, and the human immune system; classes and global health advocacy work had taught me about the human side of the epidemic, but living in Cambridge, MA, I remained a world away.

Through a series of fortunate events, I found myself on a Skype call with Dr. William Bishai, a TB researcher from Johns Hopkins who had recently been named the first permanent director of K-RITH. His enthusiasm for basic science and the potential impact that K-RITH could have on the surrounding community was infectious, and a few months later, I was boarding a flight for Durban. My work was going to be supported by a Fulbright grant, which I had applied to with a proposal for doing research in a space I couldn't imagine and that actually didn't even exist yet. I had absolutely no idea what to expect from K-RITH, and was terrified that I was proposing an idea that was doomed to fail. This uncertainty wasn't limited to the feasibility of my project; by the time I was headed to South Africa, I had signed a lease for an apartment thousands of miles away that I had never seen, and knew I would have to buy a car with a manual transmission after only a few frustrating lessons in driving stick shift at home.

With my degree in hand and an idealistic dream of making some groundbreaking contribution to TB research in mind, I arrived amidst the chaos of the official opening of the state-of-the-art research facility that had been drawn up by Dr. Walker and the K-RITH steering committee in 2009. The shiny new building quickly filled with brand new laboratory equipment, brilliant scientists from all over the world, precious samples from patients suffering from TB and HIV infection, and a few idealistic recent college grads like me. Between the smart people, the millions of dollars, and the myriad fascinating scientific questions to be answered, it seemed like all of the ingredients were in place for realizing the ambitious vision born six years prior at HHMI.

After the shaky first months moving equipment into the brand new lab spaces, we settled into our new home. At this point, I still hadn't started the project I had proposed to do at K-RITH—studying the immune system of patients co-infected with HIV and TB. Though I kept myself busy with side projects in the meantime, I was anxious to do the project I had envisioned working on back in 2011. My project was slow to start because it took months for the lab to receive reagents shipped from overseas. Certain products manufactured or stocked by companies with South African offices didn't pose a significant problem, but some experiments required products that were shipped from the US or Europe, and, in one case, took months to arrive. This delay made the cost of a failed experiment much higher than I had experienced before, and required that both time and materials be used as efficiently as possible.

When I did start my research, it was with a very different sense of purpose than the research I did as an undergraduate. Whereas in Boston I studied cells purchased from a company and stored in the freezer, at K-RITH I was working with blood and tissue that came directly from the patients that were being treated nearby.

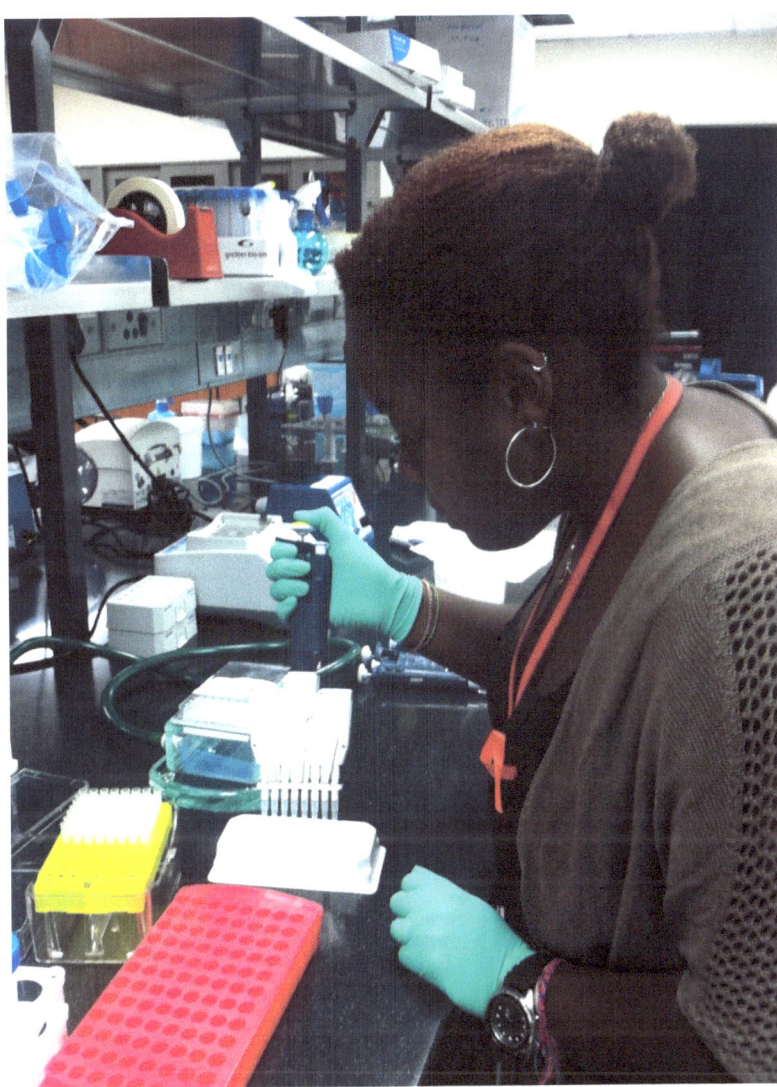

The author, at
work at the bench
at K-RITH.

Suddenly 'TB' and 'HIV' represented much more than stocks of bacteria and virus with which I infected cultured cells in the lab in Boston. Living at the heart of these epidemics, shadowing infectious disease doctors at the country's second largest public hospital, the human side of these epidemics was devastatingly apparent. As an American researcher in a well-funded lab, I was insulated from the worst of the tragedy, but even inside the lab we saw the effects the disease we were studying first-hand: a close friend and labmate struggled for two years with nearly fatal battle with MDR-TB. His resilience was an inspiration to all of us, and the frankness with which he shared the narrative of his illness incarnated the narratives of the 2006 XDR-TB

crisis that inspired K-RITH's mission.

Carrying out this mission has further challenges going forward. According to Dr. Walker, the foremost challenge will be sustainability. While the problem of financial stability is not unique to K-RITH or to South Africa, the politics of funding research for diseases that affect the developing world complicate the situation. NIH funding comes from American taxpayer dollars, and the continued funding of research programs that affect people outside of the US requires continued taxpayer support. Would Americans be willing to fund research on diseases that don't affect them directly? To fund TB research, other organizations have stepped up, such as The Bill and Melinda Gates Foundation, and others that, fortunately for K-RITH, specifically fund projects in South Africa.

As a scientist, I'm used to challenges with equipment, reagents, and funding. But I didn't expect that one of the biggest challenges of working in South Africa would be personal. Daily, I found myself emotionally exhausted by the struggles of living as a black American woman in a South Africa that was still licking its post-apartheid wounds. Sometimes this meant artfully maneuvering my way through variably subtle but unmistakably racist comments and situations, a skill that I had long since mastered growing up in the United States, but nevertheless found fatiguing. Other times it meant grappling with my identity as I balanced the novelty of being in the country's racial majority with being constantly reminded that I was distinctly and irrefutably foreign. And other times it meant facing the sobering reality, as I catalogued the demographics of various HIV patient cohorts at K-RITH, that the brunt of the HIV epidemic was borne by young, black women like myself that happened to have been born into less privileged circumstances.

Overall, the experience was demoralizing and isolating, and several months into my first year

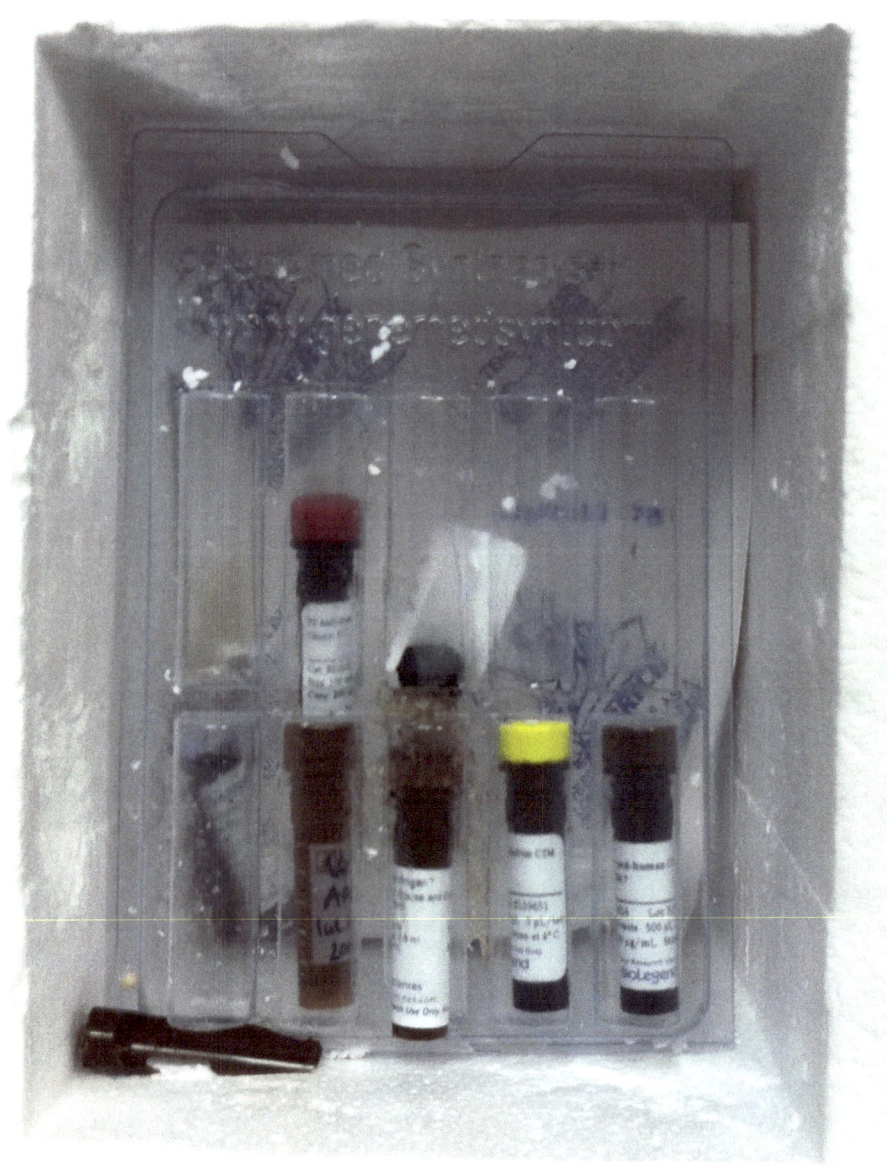

Vials with precious samples were shipped from the US to the lab at the wrong temperature, causing some of the tubes to burst and holding back research.

at K-RITH, part of me was looking forward to going back to the US on my scheduled departure date in June 2013. But in February, South African scientists made history when a team of researchers at the University of Cape Town announced the results of the first TB vaccine to be tested in humans in 92 years. Though the results of the trial were disappointing, its very existence signified a historic achievement for the global TB community, and the South African community in particular. The very existence of this trial illustrated the indispensability of places like K-RITH—addressing diseases that affect the developing world requires that the populations affected by them be central players in discovering their cures. Appreciating that K-RITH afforded me the opportunity to do research I could not do anywhere else in the world, I decided to stay for an additional year beyond the tenure of my Fulbright scholarship.

Representatives from the team that conducted the vaccine trial attended the 2013 Keystone TB conference in Whistler, British Columbia, which I attended with over a dozen researchers from K-RITH. This was my first time being at a big international research conference, and I felt proud to be presenting my work, but even more so to be representing K-RITH and the South African research community. The energy at this conference mirrored that of the K-RITH opening six months prior, but by this point, we were presenting exciting data, not just hope. In the posters, abstracts, and lectures delivered by my colleagues, the realization of K-RITH's potential was suddenly apparent. The historic Cape Town vaccine trial illustrated the uniquely impactful research that could be conducted in a TB/HIV-endemic setting. Eight years after the devastating XDR-TB outbreak in Tugela Ferry, South African research institutions were making groundbreaking contributions to the global fight to end the HIV and TB epidemics.

The fight continues. For me, it involves working towards a PhD degree studying the basic biology of the immune system in order to identify defense mechanisms that may play a role in the pathogenesis of TB. For the community as a whole, the problem is substantially more daunting. In 2013, 9 million people developed TB, 13 percent of whom were HIV-positive. 1.5 million people died of TB

disease in 2013, and Africa suffered the highest death rates relative to population of any region in the world. In the same year, 1.5 million people died from AIDS, and TB was the leading cause of death amongst the 35 million worldwide living with HIV. Numbers on a page don't do the epidemic justice, and researchers all over sub-Saharan Africa work tirelessly on its front lines every day. Armed with brilliant investigators, a commitment of over $70 million from HHMI, and a state-of-the-art research facility, the future of K-RITH seems promising.

My short scientific career, particularly at K-RITH, has taught me a lot about the distinction between factors that are necessary, versus those that are sufficient, to cause a certain outcome. This is something that we test in the lab experimentally when we're trying to understand the underlying mechanisms of a complex biological process. But it's also a principle that I've come to appreciate with respect to combating complex diseases like HIV and TB. Effective drugs are necessary, but insufficient, because the pathogenesis of these infections is as socially determined as it is biological. Similarly, establishing a successful research institution in South Africa to study these diseases is complicated by factors that are both tangible and abstract, and affect the institution's health on both individual and systemic levels. Financial stability, talented scientists, and cutting-edge technology; like effective drugs and vaccines for diseases, these ingredients are necessary constituents of a successful international research facility. Unfortunately, like drugs and vaccines, they're also not sufficient. And like the intangible and often under-appreciated social determinants of diseases, the more abstract building blocks of K-RITH's success are much harder to identify and address. Just as social interventions have been effective at curbing the worldwide incidence of HIV and TB in recent decades, K-RITH's institutional mindfulness of its less perceptible challenges will be integral to its continued success. ■

Zuri Sullivan is a graduate student in the Department of Immunobiology at Yale University, and a blogger and founding editor of Because Science.

www.ingramcontent.com/pod-product-compliance
Lightning Source LLC
Chambersburg PA
CBHW050857180526
45159CB00007B/2709

* 9 7 8 1 5 0 8 6 0 1 8 2 1 *